Anonymous

Meteorology and Climatology of the Great Valleys and

Foothills of California

Anonymous

Meteorology and Climatology of the Great Valleys and Foothills of California

ISBN/EAN: 9783744743181

Printed in Europe, USA, Canada, Australia, Japan

Cover: Foto ©berggeist007 / pixelio.de

More available books at **www.hansebooks.com**

METEOROLOGY AND CLIMATOLOGY

OF THE

GREAT VALLEYS AND FOOTHILLS

OF

CALIFORNIA,

FOR FROM FIFTEEN TO THIRTY-SIX YEARS.

Collated and compiled by Sergeant JAMES A. BARWICK, Observer Signal Corps, U. S. A.,
and Meteorologist to the State Board of Agriculture.

SACRAMENTO:

STATE OFFICE,..........JAMES J. AYERS, SUPT. STATE PRINTING.

1886.

Compliments of

SERGEANT JAMES A. BARWICK,

Observer Signal Corps U. S. A.,

AND METEOROLOGIST TO THE STATE BOARD OF AGRICULTURE,

Sacramento, California.

[PLEASE ACKNOWLEDGE RECEIPT OF THIS REPORT.]

CONTENTS.

METEOROLOGY AND CLIMATOLOGY

OF THE

GREAT VALLEYS AND FOOTHILLS OF CALIFORNIA

FOR FROM FIFTEEN TO THIRTY-SIX YEARS.

Collated and compiled by SERGEANT JAMES A. BARWICK, Observer Signal Corps U. S. A., and Meteorologist to the State Board of Agriculture.

To the Secretary of the State Agricultural Society of California:

SIR: 1 have the honor to submit the following meteorological report on the climate of California, and more especially that of the great interior valleys of this State. There will be found the rainfall by seasons, Spring, Summer, Autumn, and Winter, for Sacramento, compiled from observations taken by Dr. T. M. Logan, Dr. F. W. Hatch, and those of the United States Signal Service, covering a period from 1853 to April 1, 1886. Also a general review of the meteorological condition of Sacramento, as deduced from Signal Service observations from July 1, 1877, to February 28, 1886; showing the pressure of the atmosphere by seasons, the temperature, direction of wind, velocity of wind, clear, fair, cloudy, and rainy days, and various other data pertaining to the climatic conditions of the above named city. A tabulated statement of rainfall by months, years, and seasons, from near Fort Jones, in Scott Valley, Yreka, Red Bluff, Oroville, Marysville, Colusa, Princeton, West Butte, Grass Valley, Placerville, Georgetown, Nicolaus, Folsom City, Sacramento, Oakland, San Francisco, Salinas, Santa Barbara, Los Angeles, San Bernardino, San Diego, and Poway; the above places give the rainfall for from one to thirty-four years, making quite a fair average estimate of the precipitation from San Diego to Siskiyou, and from the Sierras to the sea. A supplemental report of the rainfall for January, 1886, and for the season of 1885–6, up to February 1, for the above named places. Also a tabulated statement of the average rainfall for January and February for many years, and the rainfall for January and February, 1886. The average seasonal rainfall up to March first, for many years, along with the rainfall for this season, 1885–6, up to March first, for each Signal Service Station, and for the stations of the Southern Pacific Railroad Company, voluntary observers, and Post Surgeons. This data was collated and tabulated at the United States Signal Service Office, Division of the Pacific, at San Francisco, Lieut. W. A. Glassford, United States Army, assistant officer in charge.

A complete meteorological report and weather review of the climate of Oakland for 1885, and comparison for ten years past, by J. B. Trembley, M.D., Oakland.

An instructive and interesting article entitled, "Storms on the Pacific Coast of North America," from the annual report of the Chief Signal Officer of the Army.

A portion of two articles by the late the Honorable B. B. Redding, and heretofore published in the report of the Agricultural Society of California, for the years 1877 and 1878.

That published in 1877 bears more especially upon the "temperature distributions of the interior valleys of this State, and the effects the great deserts of California and Nevada have upon it;" and the one in 1878 on the "climatic condition of the entire valley from Redding on the north to Sumner on the south," in which will be found the essence, *foundation, as it were*, of the climatic conditions of the great valleys and foothills in regard to the cultivation of all kinds of fruits, both semi-tropical and otherwise; gives information how to choose, in the foothills and valley, proper sites for certain fruits, grains, etc.

A brief synopsis of the meteorological features of the following signal service stations in the order named: San Diego, Los Angeles, San Francisco, Sacramento, and Red Bluff. A table of interesting matter, composed of the most prominent places of California, and other portions of the United States, and health resorts of Europe and Mexico, showing the mean annual temperature, mean temperature for Spring, Summer, Fall, and Winter, along with the highest and lowest temperature for many years. This table shows that California compares with other noted health resorts and prominent places in a highly complimentary manner to this State.

The "North Winds of California," always an interesting study, have been reproduced, being from the pen of the Rev. J. H. C. Bonté, Secretary of the State University of California.

An excellent report on the climate of Palestine, and more particularly Jerusalem, by Mr. Selah Merrill, United States Consul at the latter place. That climate will be found to resemble our own in a great many respects, especially at Colfax, which is nearly the same elevation as Jerusalem. It is an article that will well repay a careful perusal.

The meteorological report of California for 1885 embodies all the most salient points of the climate of this State that has been written and published heretofore at various times and places, being a gathering of all worthy articles known to the writer on the climatic peculiarities of this State, making this report (1885) of especial value as a book of reference on subjects relating to various conditions of a meteorological and climatological nature.

Very respectfully, your obedient servant,

SERGEANT JAMES A. BARWICK,

Observer Signal Corps, United States Army, and Meteorologist to the State Board of Agriculture.

SACRAMENTO, Cal., February 28, 1886.

RAINFALL, TEMPERATURE, ETC., AT SACRAMENTO, CAL.

Sacramento is situated in north latitude, 38° 35′; longitude west from Greenwich, 121° 30′; above sea level, 30 feet; height of barometer cistern above sea level, 64 feet. •

RAINFALL FOR THE WINTER SEASONS.

The following tabulated statement shows the rainfall and number of days that rain fell for each Winter month; also, the total rainfall and total number of days that rain fell during the entire Winter seasons; the Winter seasons beginning with the Winter of 1852–3, and ending with the Winter of 1885–6. The three Winter months composing the Winter season that gave the heaviest rainfall was during the Winter of 1861–2: 27.94 inches. The driest Winter was that of 1863–4 : 3.09 inches:

WINTER OF—	DECEMBER. Inches.	No. of Days.	JANUARY. Inches.	No. of Days.	FEBRUARY. Inches.	No. of Days.	Total for Winter Months.	Total No. Days for Winter.
1852–3	13.41	20	3.00	12	2.00	6	18.41	38
1853–4	1.54	4	3.25	6	8.50	14	13.29	24
1854–5	1.15	8	2.67	15	3.46	7	7.28	30
1855–6	2.00	13	4.92	16	.09	6	7.61 ·	35
1856–7	2.40	13	1.38	14	4.80	17	8.58	44
1857–8	2.63	13	2.44	21	2.46	13	7.53	47
1858–9	4.34	17	.96	19	3.91	18	9.21	54
1859–60	1.83	17	2.31	15	.93	14	5.07	46
1860–1	4.28	20	2.67	10	2.92	9	9.87	39
1861–2	8.64	22	15.04	20	4.26	11	27.94	53
1862–3	2.33	11	1.73	10	2.75	11	6.81	32
1863–4	1.82	10	1.08	7	.19	2	3.09	19
1864–5	7.87	16	4.78	13	.71	6	13.36	35
1865–6	.36	9	7.70	18	2.01	11	10.07	38
1866–7	9.51	21	3.44	15	7.10	9	20.05	45
1867–8	12.85	18	6.04	17	3.15	9	22.04	44
1868–9	2.61	11	4.79	14	3.63	5	11.03	30
1869–70	1.96	7	1.37	9	3.24	11 ·	6.57	27
1870–1	.97	6	2.08	8	1.92	11	4.97	25
1871–2	10.49	22	4.04	11	4.74	18	19.47	51
1872–3	5.39	13	1.23	10	4.46	17	11.08	40
1873–4	10.01	21	5.20	14	1.86	9	17.07	44
1874–5	.44	17	8.70	14	.55	2	9.69	33
1875–6	5.52	14	4.99	13	3.75	10	14.26	37
1876–7			2.77	11	1.04	9	3.81	20
1877–8	1.43	5	9.26	17	8.04	17	18.73	39
1878–9	.47	3	3.18	11	3.88	9	7.53	23
1879–80	3.41	12	1.64	7	1.83	10	6.88	29
1880–1	11.81	21	6.14	9	5.06	13	23.01	43
1881–2	3.27	11	1.89	8	2.40	6	7.56	25
1882–3	1.13	8	2.23	4	1.11	3	4.47	· 15
1883–4	.44	6	3.43	9	4.46	10	8.33	25
1884–5	10.45	11	2.16	8	.49	6	13.10	25
1885–6	5.76	10	7.95	13	.29	5	14.00	28
Totals	152.75	430	136.46	418	102.59	334	391.77	1,182
Averages for 34 years	4.493	12.6	4.014	12.3	3.017	9.8	11.523	34.8

4

MEAN SPRING RAINFALL.

The table below will be found to contain the record of rainfall and number of days rain fell during the Spring months and for the Spring season. It informs us that the wettest Spring season was that of 1880— 16.66 inches; and the driest was that of 1857—.68 of an inch; the mean average for thirty-three years being 5.219 inches, showing a deficiency of nearly 5 inches in the dry season of 1857, and an excess of over 11 inches during the wet Spring of 1880, as compared with a thirty-three years average:

SPRING OF—	MARCH.		APRIL.		MAY.		Total for Spring Months.	Total No. Days for Spring.
	Inches.	No. of Days.	Inches.	No. of Days.	Inches.	No. of Days.		
1853	7.00	8	3.50	7	1.45	4	11.95	19
1854	3.25	4	1.50	9	.21	4	4.96	17
1855	4.20	9	4.32	9	1.15	6	9.67	24
1856	1.40	5	2.13	8	1.84	4	5.37	17
1857	.68	10	sprink.	1	sprink.	3	.68	14
1858	2.88	13	1.21	3	.20	4	4.29	20
1859	1.64	14	.98	6	1.04	4	3.66	24
1860	5.11	17	2.87	8	2.49	10	10.47	35
1861	3.32	7	.48	4	.59	3	4.39	14
1862	2.80	15	.82	9	1.81	9	5.43	33
1863	2.36	10	1.69	9	.36	2	4.41	21
1864	1.39	12	1.08	4	.74	8	3.12	24
1865	.48	7	1.37	3	.46	2	2.31	12
1866	2.02	11	.48	6	2.25	5	4.75	22
1867	1.01	6	1.80	7	.01	1	2.82	14
1868	4.35	12	2.31	9	.27	2	6.93	23
1869	2.94	12	1.24	5	.65	2	4.83	19
1870	1.64	6	2.12	7	.27	1	4.03	14
1871	.69	8	1.45	6	.76	5	2.90	19
1872	1.94	10	.61	6	.28	3	2.83	19
1873	.55	4	.51	4			1.06	8
1874	3.05	10	.89	10	.37	6	4.31	26
1875	.80	9	sprink.	3	sprink.	1	.80	13
1876	4.15	13	1.10	10	.15	4	5.40	27
1877	.56	7	.19	7	.64	6	1.39	20
1878	3.09	14	1.07	3	.17	4	4.33	21
1879	4.88	14	2.66	12	1.30	5	8.84	31
1880	1.70	7	14.20	15	.76	3	16.66	25
1881	1.37	6	1.64	6	sprink.	1	3.01	13
1882	3.78	10	1.99	8	.35	1	6.12	19
1883	3.70	6	.67	7	2.85	9	7.22	22
1884	8.14	13	4.32	9	.06	3	12.52	25
1885	.08	2	.08	7	sprink.	1	.76	10
Totals	86.86	311	61.88	227	23.48	126	172.22	664
Averages for 33 years	2.632	9.4	1.875	6.9	.711	3.8	5.219	20.1

MEAN SUMMER RAINFALL.

In the recorded statement below will be found the rainfall for each month of our dry Summer season; also, the total for the season, as well as the total number of days, etc., that rain fell. The average for the thirty-three years past is .168 of an inch. The Summer season that gave the most rainfall was that of 1884—1.45 inches. But five seasons of the thirty-three gave none, those being 1859, 1863, 1867, 1878, and 1883:

SUMMER OF—	JUNE.		JULY.		AUGUST.		Total for Summer Months.	Total No. Days for Summer.
	Inches.	No. of Days.	Inches.	No. of Days.	Inches.	No. of Days.		
1853	sprink.	1	sprink.	2			sprink.	3
1854	.31	2			sprink.	1	.31	3
1855	.01	1					.01	1
1856	.03	1					.03	1
1857	.35	2			sprink.	1	.35	3
1858	.10	2	.01	1	sprink.	4	.11	7
1859								
1860	.02	2	.03				.05	2
1861	.14	4	.55	3			.69	7
1862	.01	1			.01	1	.02	2
1863								
1864	.09	3			.08	6	.17	9
1865			sprink.	3			sprink.	3
1866	.10	2	.02	3			.12	5
1867								
1868	sprink.	3					sprink.	3
1869	.01	1					.01	1
1870	sprink.	1	sprink.	1	sprink.	1	sprink.	3
1871	sprink.	1					sprink.	1
1872	.02	1					.02	1
1873	sprink.	1	.02	2	sprink.	1	.02	4
1874	sprink.	2	sprink.	1			sprink.	3
1875	1.10	2					1.10	2
1876			.21	2	.02	1	.23	3
1877	.01	1	sprink.	1	sprink.	1	.01	3
1878								
1879	.13	1	sprink.	1	sprink.	1	.13	3
1880			sprink.	1			sprink.	1
1881	.50	2	sprink.	1			.50	3
1882	.10	1	sprink.	1			.10	2
1883								
1884	1.45	7			sprink.	1	1.45	8
1885	.11	2	sprink.	1	none		.11	3
Totals	4.59	47	.84	24	.11	19	5.54	90
Averages for 33 years.	.139	1.4	.026	0.7	.003	0.6	.168	2.7

MEAN AUTUMNAL RAINFALL.

The table following shows the Autumnal rainfall by months and total for the Fall season, both of rainfall and number of days rain fell, for the last thirty-three years, the wettest being the Fall season of 1885—11.44 inches; the driest being that of 1880—.05 of an inch:

FALL OF—	SEPTEMBER.		OCTOBER.		NOVEMBER.		Total for Autumn Months.	Total No. Days for Autumn.
	Inches.	No. of Days.	Inches.	No. of Days.	Inches.	No. of Days.		
1853	sprink.	1	sprink.	1	1.50	5	1.50	7
1854	sprink.	1	1.01	11	.65	2	1.66	14
1855	sprink.	1			.75	9	.75	10
1856	sprink.	1	.20	6	.65	10	.85	17
1857			.65	3	2.41	10	3.06	13
1858	sprink.	5	3.01	5	.15	11	3.16	21
1859	.02	3			6.48	15	6.50	18
1860	.06	2	.91	9	.18	5	1.15	16
1861			sprink.	1	2.17	12	2.17	13
1862			.36	6	sprink.	2	.36	8
1863	sprink.	1			1.49	7	1.49	8
1864	sprink.	1	.12	2	6.72	9	6.84	12
1865	.08	4	.48	5	2.43	9	2.99	18
1866			sprink.	1	2.43	8	2.43	9
1867	.01	1			3.81	6	3.82	7
1868					.77	5	.77	5
1869	sprink.	1	2.12	2	.85	5	2.97	8
1870			.02	2	.58	6	.60	8
1871	sprink.	1	.21	1	1.22	8	1.43	10
1872	sprink.	2	.22	2	1.93	4	2.15	8
1873			.31	4	1.21	5	1.52	9
1874	.05	1	2.26	8	3.80	9	6.11	18
1875			.44	4	.6.20	10	6.64	14
1876	sprink.	1	3.45	7	.30	1	3.75	9
1877			.73	5	1.07	7	1.80	12
1878	.29	3	.55	1	.51	3	1.35	7
1879			.88	4	2.05	8	2.93	12
1880					.05	2	.05	2
1881	.30	1	.55	6	1.88	4	2.73	11
1882	.57	2	2.63	6	3.22	7	6.42	15
1883	.90	2	.97	6	.61	3	2.48	11
1884	.60	3	2.01	4			2.61	7
1885	.08	1	.02	2	11.34	17	11.44	20
Totals	2.96	39	24.11	114	69.41	224	96.48	377
Averages for 33 years.	.090	1.2	.731	3.5	2.104	6.8	2.924	11.4

YEARLY AND SEASONAL RAINFALL, ETC.

The instructive tabulated information below gives the rainfall annually—that is, from January to December of each year—for thirty-three years. Also, the rainfall by seasons, beginning with September first of one year and ending with August thirty-first of the next year, the wettest season being 1861-2—35.56 inches; the driest that of 1863-4—7.87 inches; the wettest calendar year being 1880—31.99 inches; the driest being 1877—8.44 inches; the mean average seasonal rainfall for thirty-two years being 19.076 inches; the mean average for the year, or the mean annual average, being 19.529 inches. The difference between the mean average rainfall, calculating from January first to December thirty-first of each year, and from September first of one year to August thirty-first of next year, is .453 of an inch in favor of the calendar year:

YEAR OF—	Yearly Rainfall.	Total No. of Days Rain Fell.	Season of—	Rainfall— Inches.	Total No. of Days.
1853	19.99	51			
1854	19.83	62	1853–54	20.06	51
1855	18.56	70	1854–55	18.62	69
1856	14.26	70	1855–56	13.76	63
1857	12.01	74	1856–57	10.46	78
1858	16.80	99	1857–58	15.00	87
1859	16.86	97	1858–59	16.03	100
1860	19.19	72	1859–60	22.09	101
1861	21.38	75	1860–61	16.10	76
1862	27.44	85	1861–62	35.56	100
1863	12.20	60	1862–63	11.58	64
1864	19.27	67	1863–64	7.87	57
1865	11.15	61	1864–65	22.51	62
1866	26.52	86	1865–66	17.93	83
1867	30.03	63	1866–67	25.30	72
1868	19.50	68	1867–68	32.79	78
1869	18.19	52	1868–69	16.64	58
1870	10.21	51	1869–70	13.57	52
1871	19.32	71	1870–71	8.47	53
1872	19.17	70	1871–72	24.05	83
1873	18.20	69	1872–73	14.21	60
1874	17.92	87	1873–74	22.90	82
1875	23.31	59	1874–75	17.70	71
1876	18.12	62	1875–76	26.53	75
1877	8.44	60	1876–77	8.96	54
1878	23.45	65	1877–78	24.86	72
1879	22.37	78	1878–79	17.85	64
1880	31.99	66	1879–80	26.47	67
1881	20.71	60	1880–81	26.57	61
1882	18.06	58	1881–82	16.51	57
1883	13.48	46	1882–83	18.11	52
1884	34.92	70	1883–84	24.78	68
1885	20.72	57	1884–85	16.58	58
			1885–86	‡28.12	53
Totals	644.47	2.241		610.42	2.228
Averages	19.520	*67.9		†19.076	69.6

* Mean for thirty-three years.
† Mean for thirty-two seasons.
‡ Up to April 1, 1886.

RAINFALL FOR SPRING, SUMMER, AUTUMN, WINTER, AND TOTAL FOR EACH YEAR.

The following table gives the rainfall for each season of Spring, Summer, Autumn, and Winter; also the total rainfall for each year. The table shows the annual rainfall for each year, beginning with the year 1850. The rainfall for the Winter seasons begins with the Winter of 1849–50, and ends with the Winter of 1885–86, making a total of thirty-seven Winters:

YEAR.	Rainfall for Spring.	Rainfall for Summer.	Rainfall for Autumn.	Rainfall for Winter.	Annual Rainfall.
1849			4.00		*16.50
1850	14.50	none	sprinkle	17.80	19.50
1851	3.71	none	3.32	1.00	15.10
1852	6.89	none	6.00	7.77	27.00
1853	11.95	sprinkle	1.50	18.41	19.99
1854	4.96	.31	1.66	13.29	19.83
1855	9.67	.01	.75	7.28	18.56
1856	5.37	.03	.85	7.61	14.26
1857	.68	.35	3.06	8.58	12.91
1858	4.29	.11	3.16	7.53	16.80
1859	3.66	none	6.50	9.21	16.86
1860	10.47	.05	1.15	5.07	19.19
1861	4.30	.69	2.17	9.87	21.38
1862	5.43	.02	.36	27.94	27.44
1863	4.41	none	1.49	6.81	12.20
1864	3.12	.17	6.84	3.09	19.27
1865	2.31	sprinkle	2.99	13.36	11.15
1866	4.75	.12	2.43	10.07	26.52
1867	2.82	none	3.82	20.05	30.03
1868	6.93	sprinkle	.77	22.04	19.50
1869	4.83	.01	2.97	11.03	18.19
1870	4.03	sprinkle	.60	6.57	10.21
1871	2.90	sprinkle	1.43	4.97	19.32
1872	2.83	.02	2.15	19.47	19.17
1873	1.06	.02	1.52	11.08	18.20
1874	4.31	sprinkle	6.11	17.07	17.92
1875	.80	1.10	6.64	9.69	23.31
1876	5.40	.23	3.75	14.26	18.12
1877	1.39	.01	1.80	3.81	8.44
1878	4.33	none	1.35	18.73	23.45
1879	8.84	.13	2.93	7.53	22.37
1880	16.66	sprinkle	.05	6.88	31.99
1881	3.01	.50	2.73	23.01	20.71
1882	6.12	.10	6.42	7.56	18.06
1883	7.22	none	2.48	4.47	13.48
1884	12.52	1.45	2.61	8.33	34.92
1885	.76	.11	11.44	13.10	20.72
1886				14.00	
Totals	197.32	5.54	109.80	418.04	707.07
Average	5.481	.154	2.968	11.298	19.641

*Rainfall for September, October, November, and December, 1849.

MEAN WINTER TEMPERATURE.

The tabulated statement below shows the average temperature by months, and for the season also. The Winter seasons, beginning with the season of 1853–4, and ending with the one of 1885–6; also, showing a mean average for thirty-three years. Judging from the average temperature for each season, we must conclude that the season of 1879–80 was the coldest— 44.5°; the warmest being the season of 1881—51.0°; the mean average of thirty-two years being 48.3°:

WINTER SEASON OF—	Mean Temp.— December.	Mean Temp.– January.	Mean Temp.— February.	Mean Winter Temperature.
1853–54	48.0	43.0	51.0	47.3
1854–55	47.9	43.7	52.5	48.0
1855–56	46.0	48.0	52.6	48.9
1856–57	43.9	48.5	50.2	47.5
1857–58	47.4	45.0	52.2	48.2
1858–59	44.5	44.9	50.5	46.6
1859–60	43.5	46.2	49.8	46.5
1860–61	49.3	47.1	52.2	49.5
1861–62	50.9	46.4	47.5	48.3
1862–63	46.4	46.9	48.0	47.1
1863–64	46.5	49.2	53.6	49.8
1864–65	50.2	47.4	49.0	48.9
1865–66	44.1	46.5	63.5	51.4
1866–67	50.2	48.2	47.8	48.7
1867–68	46.8	47.0	50.5	48.1
1868–69	47.0	47.6	49.9	48.2
1869–70	46.5	48.6	51.1	48.7
1870–71	45.5	48.3	49.4	47.7
1871–72	48.7	48.5	53.3	50.2
1872–73	49.0	52.7	48.2	50.0
1873–74	47.7	45.7	49.3	47.6
1874–75	45.0	46.9	52.7	48.2
1875–76	48.0	48.8	50.2	49.0
1876–77	45.5	49.1	55.0	49.9
1877–78	48.6	49.7	51.3	49.9
1878–79	47.2	45.5	55.0	49.2
1879–80	44.0	43.5	46.0	44.5
1880–81	50.3	49.2	53.5	51.0
1881–82	46.2	45.1	46.3	45.9
1882–83	48.2	41.9	46.0	45.4
1883–84	44.2	46.6	46.9	45.9
1884–85	48.8	47.1	54.0	50.0
1885–86	49.1	45.7	53.3	49.4
Totals	1555.1	1548.5	1682.3	1595.5
Averages for 33 years	47.1	46.9	51.0	48.3

MEAN SPRING TEMPERATURE.

The table below will be found to contain the average temperature by months for the Spring, also for the season. The warmest one, as indicated by its average temperature, was 1853—62.9°; the coldest, 1880—55.0°: the mean average Spring temperature being 59.5°:

SPRING SEASON OF—	Mean Temp.— March.	Mean Temp.— April.	Mean Temp.— May.	Mean Spring Temperature.
1853	59.8	61.0	68.0	62.9
1854	53.0	60.0	62.0	58.3
1855	54.8	58.1	·60.2	57.7
1856	57.0	58.8	63.9	59.9
1857	56.4	63.3	65.5	61.7
1858	53.7	59.8	65.2	59.6
1859	51.5	57.1	63.0	57.2
1860	53.3	57.8	58.5	56.5
1861	55.0	60.6	63.7	59.8
1862	53.6	58.0	61.2	57.6
1863	57.6	59.5	67.1	61.4
1864	56.1	62.1	68.5	62.2
1865	53.6	59.3	70.2	61.0
1866	54.2	61.9	63.1	59.7
1867	50.7	59.7	64.4	58.3
1868	55.0	60.1	64.2	59.8
1869	53.6	59.0	64.2	58.9
1870	53.0	57.0	61.0	57.0
1871	56.0	59.2	61.5	58.9
1872	56.8	57.6	67.0	60.5
1873	56.8	60.0	67.9	61.6
1874	52.9	59.5	64.7	59.0
1875	58.7	63.0	68.1	63.3
1876	54.6	59.5	65.7	59.9
1877	50.0	60.2	64.5	61.2
1878	56.7	59.4	65.5	60.5
1879	57.4	60.3·	60.2	59.3
1880	48.8	54.6	61.6	55.0
1881	55.5˙	60.8	64.8	60.4
1882	53.0	55.8	64.0	57.6
1883	56.9	56.0	62.6	58.5
1884	52.9	56.7	64.0	57.9
1885	59.1	60.6	65.7	61.8
Totals	1817.0	1956.3	2121.7	1964.9
Averages for 33 years	55.1	59.3	64.3	59.5

11

MEAN SUMMER TEMPERATURE.

The average temperature in the following table is for the Summer months and for the Summer season, showing by their average temperature that 1866 was the warmest—74.8°; and the coldest to have been 1880—69.1°; the mean average for thirty-three years is 71.7°; the season of 1866 being 2.9° above the mean average, and 1880 being 2.6° below the mean average for the past thirty-three years:

SUMMER SEASON OF—	Mean Temp.—June.	Mean Temp.—July.	Mean Temp.—August.	Mean Summer Temperature.
1853	77.0	75.0	71.0	74.3
1854	67.0	80.6	69.5	72.4
1855	71.1	72.5	73.0	72.2
1856	71.1	75.1	69.6	71.9
1857	71.9	71.4	71.3	71.5
1858	69.4	70.8	70.6	70.3
1859	74.8	69.1	67.2	70.4
1860	65.6	73.2	73.5	70.8
1861	66.2	73.6	69.7	69.8
1862	69.3	73.2	75.0	72.5
1863	69.1	75.6	70.7	71.8
1864	71.1	74.8	74.7	73.5
1865	73.5	74.0	71.7	73.1
1866	72.2	76.2	76.0	74.8
1867	70.3	73.7	71.7	71.9
1868	69.5	73.8	71.2	71.5
1869	70.8	74.3	71.3	72.1
1870	69.3	71.8	72.6	71.2
1871	70.1	70.2	72.0	70.8
1872	69.2	71.4	73.1	71.6
1873	71.7	73.2	66.3	70.4
1874	70.2	72.8	70.9	71.3
1875	70.6	73.3	72.5	72.1
1876	76.9	74.0	72.8	74.6
1877	72.5	75.0	72.9	73.5
1878	71.8	73.4	73.4	72.9
1879	72.1	71.8	74.7	72.9
1880	66.6	70.9	69.7	69.1
1881	66.0	71.1	68.2	68 5
1882	68.1	73.4	71.9	71.1
1883	72.6	73.1	71.4	72.4
1884	65.8	71.2	72.5	69.8
1885	66.2	71.0	73.0	70.1
Totals	2319.6	2414.5	2365.6	2366.6
Averages for 33 years	70.3	73.2	71.7	71.7

MEAN AUTUMN TEMPERATURE.

The average temperature for the Fall season indicates the Fall of 1853 as being the warmest, 69.0°; that of 1881 was the coldest, judging from the average temperature, 58.5°. The average mean temperature for thirty-three years past, 61.5°, showing the average of 1853 to have been 7.5° above the mean average, and that of 1881 to have been 3.0° below the mean average temperature for the past thirty-three years:

FALL SEASON OF—	Mean Temp.—September.	Mean Temp.—October.	Mean Temp.—November.	Mean Autumn Temperature.
1853	76.0	78.0	53.0	69.0
1854	65.0	60.0	55.0	60.0
1855	68.0	63.0	50.6	60.5
1856	70.9	58.0	52.2	60.4
1857	67.9	61.5	53.2	60.9
1858	68.9	59.5	54.2	60.9
1859	65.9	63.3	54.0	61.1
1860	67.6	59.8	53.5	60.3
1861	67.8	59.9	53.6	60.4
1862	70.4	67.6	53.1	63.7
1863	69.0	62.8	52.7	61.5
1864	69.8	64.5	53.5	62.6
1865	68.8	63.1	56.9	62.9
1866	72.2	65.2	53.8	63.7
1867	68.8	62.7	54.8	62.1
1868	68.3	62.0	53.9	61.4
1869	69.9	63.1	54.0	62.3
1870	68.0	63.6	53.4	61.7
1871	67.4	62.2	50.2	59.9
1872	68.8	58.9	51.2	59.6
1873	69.9	61.4	57.5	62.9
1874	70.7	61.7	53.9	62.1
1875	55.7	69.9	56.7	60.8
1876	70.1	63.5	53.3	62.3
1877	72.7	62.9	54.7	63.4
1878	69.0	62.9	55.5	62.5
1879	70.4	61.5	50.9	60.9
1880	68.0	62.1	49.7	59.9
1881	67.8	56.8	50.8	58.5
1882	68.4	58.4	49.5	58.8
1883	71.6	58.2	50.5	60.1
1884	64.8	59.9	55.3	60.0
1885	69.8	64.3	54.4	62.8
Totals	2268.3	2062.2	1759.5	2029.9
Averages for 33 years	68.7	62.5	53.3	61.5

AVERAGE ANNUAL AND SEASONAL TEMPERATURES.

The statement below shows the average temperature, for each year, for thirty-three years; for the Spring, Summer, and Autumn, for thirty-three years, and the average Winter temperature for thirty-two years. The coldest year, inferring from the average temperature, was that of 1880—57.5°; the warmest was 1864—62.8°; the mean average for the past thirty-three years being 60.2°, showing the coldest to have been 2.7° below the mean average, while the warmest year being that of 1864, when it was 2.6° above the mean average for thirty-three years. By careful study of the following table, one is struck by the slight difference between the coldest and warmest year, as compared with a thirty-three years average, generally not more than 3°. That is, we might safely say that the average temperature of any year is not likely to vary more than 3° from 60°, either way, between the hottest and coldest year, as compared with the mean average temperature for the past thirty-three years:

YEAR.	Mean Annual Temperature.	Mean Spring Temperature.	Mean Summer Temperature.	Mean Autumn Temperature.	Mean Winter Temperature.
1853	62.6	62.9	74.3	69.0	*.........
1854	59.5	58.3	72.4	60.0	47.3
1855	59.5	57.7	72.2	60.5	48.0
1856	60.1	59.9	71.9	60.4	48.9
1857	60.7	61.7	71.5	60.9	47.5
1858	59.5	59.6	70.3	60.9	48.2
1859	58.7	57.2	70.4	61.1	46.6
1860	59.0	56.5	70.8	60.3	46.5
1861	60.1	59.8	69.8	60.4	49.5
1862	62.2	57.6	72.5	63.7	48.3
1863	60.3	61.4	71.8	61.5	47.1
1864	62.8	62.2	73.5	62.6	49.8
1865	61.0	61.0	73.1	62.9	48.9
1866	62.1	59.7	74.8	63.7	51.4
1867	59.9	58.3	71.9	62.1	48.7
1868	60.1	59.8	71.5	61.4	48.1
1869	60.4	58.9	72.1	62.3	48.2
1870	59.6	57.0	71.2	61.7	48.7
1871	59.6	58.9	70.8	59.9	47.7
1872	60.4	60.5	71.6	59.6	50.2
1873	60.7	61.6	70.4	62.9	50.0
1874	59.8	59.0	71.3	62.1	47.6
1875	62.5	63.3	72.1	60.8	48.2
1876	61.7	59.9	74.6	62.3	49.0
1877	61.2	61.2	73.5	63.4	49.9
1878	61.3	60.5	72.9	62.5	49.9
1879	60.3	59.3	72.9	60.9	49.2
1880	57.5	55.0	69.1	59.9	44.5
1881	59.2	60.4	68.5	58.5	51.0
1882	58.5	57.6	71.1	58.8	45.9
1883	58.8	58.5	72.4	60.1	45.4
1884	58.8	57.9	69.8	60.0	45.9
1885	61.2	61.8	70.1	62.8	50.0
1886					49.4
Totals	1987.6	1964.9	2365.7	2029.9	1595.5
Averages	†60.2	59.5	71.7	61.5	‡48.3

* The Winter tables are for the Winters from 1852-3 to 1885-6, both inclusive.
† Mean for thirty-three years.
‡ Mean for thirty-three years.

3

14

The following table gives the average annual barometer, thermometer, and hygrometer, the maximum and minimum temperature. The hygrometrical observations is the amount of moisture or relative humidity contained in the atmosphere; also the mean average for eight years:

YEAR.	Mean Annual Barometer.	Mean Annual Relative Humidity.	Mean Annual Temperature.	Max. Yearly Temperature.	Min. Yearly Temperature.
1878	29.946	62.2	61.3	100.5—Aug.	23.5—Dec.
1879	29.998	65.7	60.3	103.0—Aug.	25.0—Dec.
1880	30.025	64.6	57.7	98.0—July.	25.0—Jan.
1881	30.026	66.7	59.2	98.6—July.	31.9—Dec.
1882	30.030	66.0	58.2	99.8—Aug.	27.0—Dec.
1883	30.034	69.0	58.8	103.5—July.	22.0 { Jan. Feb.
1884	29.985	70.7	58.8	100.0—Aug.	21.0—Feb.
1885	29.982	67.8	61.2	105.0—Aug.	34.2—Jan.
Totals	240.026	532.7	475.5	Highest, 105.0—1885.	Lowest, 21.0—1884.
Averages	30.003	66.6	59.5	August.	February.

The Winter tables following this statement show the average Winter pressure, temperature, and relative humidity, the highest and lowest temperature for the Winter seasons of 1877-8 to 1884-5, and the mean average for nine years:

WINTER OF—	Mean Winter Barometer.	Mean Winter Relative Humidity.	Mean Winter Temperature.	Max. Winter Temperature.	Min. Winter Temperature.
1877-78	30.004	77.1	49.9	67.0—Dec.	27.0—Jan.
1878-79	30.120	68.3	49.2	73.5—Feb.	23.5—Dec.
1879-80	30.163	77.2	44.5	64.0—Feb.	25.0 { Dec. Jan.
1880-81	30.116	84.0	51.0	67.0—Feb.	35.0—Jan.
1881-82	30.169	76.4	45.9	62.8—Feb.	29.0—Dec.
1882-83	30.189	77.9	45.4	71.7—Feb.	22.0 { Jan. Feb.
1883-84	30.120	83.0	45.9	71.0—Feb.	21.0—Feb.
1884-85	30.094	77.7	50.0	70.0—Feb.	27.0—Dec.
1885-86	30.104	87.1	49.4	72.7—Feb.	27.5—Jan.
Totals	271.079	708.7	431.2	Highest, 73.5—1879.	Lowest, 21.0—1884.
Averages	30.120	78.8	47.9	February.	February.

The average Spring pressure, relative humidity and temperature, the maximum and minimum Spring temperature, also the mean average for eight years past, will be found as follows:

Spring of—	Mean Spring Barometer.	Mean Spring Relative Humidity.	Mean Spring Temperature.	Max. Spring Temperature.	'Min. Spring Temperature.
1878	29.936	67.1	60.5	91.0—May.	40.0—March.
1879	30.046	68.4	59.3	91.0—May.	38.0—March.
1880	30.031	66.2	55.0	86.0—May.	29.0—March.
1881	30.003	. 68.4	60.4	88.8—May.	37.0—March.
1882	30.037	61.9	57.6	94.6—May.	34.1—March.
1883	30.009	68.9	58.5	98.0—May.	39.8—April.
1884	29.918	73.3	57.9	85.0—May.	39.0—March.
1885	29.970	64.9	61.8	98.0—May.	39.0—April.
Totals	240.033	539.1	471.0	Highest, 98.0—1883, and 1885, in May.	Lowest, 29.0—1880. March.
Averages	30.004	67.4	58.9		

The tabulated statement following indicates the average Summer pressure, relative humidity and temperature, the maximum and minimum Summer temperature, and the mean average for eight years:

Summer of—	Mean Summer Barometer.	Mean Summer Rel. Humidity.	Mean Summer Temperature.	Max. Summer Temperature.	Min. Summer Temperature.
1878	29.817	54.7	72.9	100.5—Aug.	49.0—June.
1879	29.821	52.7	72.9	103.0—Aug.	51.0—July.
1880	29.890	59.3	69.1	98.0—July	49.0—Aug.
1881	29.903	56.3	68.5	98.6—July	48.0—June.
1882	29.898	57.0	71.1	99.8—Aug.	51.2—June.
1883	29.908	58.4	72.4	103.5—July	49.8—June.
1884	29.919	63.3	69.8	100.0—Aug.	52.9—June.
1885	29.870	55.8	70.1	105.0—Aug.	51.5—June.
Totals	239.016	457.5	566.8	Highest, 105.0—1885. August.	Lowest, 48.0—1881. June.
Averages	29.877	57.2	70.8		

The following table shows the average Autumn pressure, relative humidity and temperature, the maximum and minimum Fall temperature, and the mean averages for the past nine years:

Fall of—	Mean Fall Barometer.	Mean Fall Rel. Humidity.	Mean Fall Temperature.	Max. Fall Temperature.	Min. Fall Temperature.
1877	29.973	54.3	63.4	88.0—Sept.	37.0—Nov.
1878	29.991	54.4	62.5	92.0—Sept.	34.0—Nov.
1879	30.000	65.2	60.9	96.0—Sept.	33.0—Nov.
1880	30.035	54.9	59.9	92.0—Sept.	27.0—Nov.
1881	30.026	58.4	58.5	96.0—Sept.	32.0—Nov.
1882	30.024	69.6	58.8	99.6—Sept.	34.0—Nov.
1883	30.011	68.8	60.1	101.0—Sept.	29.0—Nov.
1884	30.000	69.1	60.0	93.5—Sept.	37.7—Nov.
1885	29.927	66.3	62.8	98.5—Sept.	38.5—Nov.
Totals	269.987	561.0	546.9	Highest, 101.0—1883. September.	Lowest, 27.0—1880. November.
Averages	29.999	62.3	60.8		

The following table will be found to contain the average direction of the wind, the total velocity, the rainfall, and the clear, fair, and cloudy days; also, days rain fell during the Winter months from 1877–8 to 1885–6:

WINTER OF—	M'n Winter Direction.	Velocity for Winter.	Rainfall for Winter.	Clear Days.	Fair Days.	Cloudy Days.	Days Rain Fell.
1877–78	S.E.	13.452	18.74	26	28	36	39
1878–79	N.	12.650	7.53	44	31	15	23
1879–80	S.E.	13.735	6.88	39	17	35	29
1880–81	S.E.	16.092	23.01	14	26	50	43
1881–82	N.	14.611	7.56	46	26	18	25
1882–83	S.E.	11.131	4.47	52	30	8	15
1883–84	S.E.	12.294	8.33	47	25	19	25
1884–85	N.W.	16.406	13.10	40	28	23	25
1885–86	N.W.	13.889	14.00	42	29	19	28
Totals		124.260	103.62	350	240	223	252
Averages	S.E.	13.807	11.51	38.9	26.7	24.8	28.0

The average direction of the wind, total velocity, the rainfall, and number of clear, fair, and cloudy days, also number of days rain fell during the Spring season, from 1878 to 1885, will be found in the following tabulated statement:

SPRING OF—	Mean Spr'g Direction.	Velocity for Spring.	Rainfall for Spring.	Clear Days.	Fair Days.	Cloudy Days.	Days Rain Fell.
1878	S.	13.962	4.33	45	28	19	21
1879	S.E.	14.530	8.84	39	34	19	31
1880	S.E.	19.653	16.66	49	24	19	25
1881	S.	14.966	3.01	60	22	10	12
1882	N.	17.774	6.12	57	19	16	19
1883	S.	15.825	7.22	54	26	12	21
1884	S.W.	18.168	12.52	46	23	23	25
1885	S.W.	16.670	.76	58	28	6	10
Totals		131.548	59.46	408	204	124	164
Averages	S.	16.444	7.432	51.0	25.5	15.5	20.5

The statistics following will be found to contain the mean direction of the wind, total velocity, the rainfall, the number of clear, fair, and cloudy days; also the number of days rain fell for the Summer season, from 1878 to 1885:

SUMMER OF—	Mean Summer Direction.	Velocity for Summer.	Rainfall for Summer.	Clear Days.	Fair Days.	Cloudy Days.	Days Rain Fell.

The mean direction of the wind, the total velocity, the rainfall, and number of clear, fair, and cloudy days; also the number of days rain fell, for the Fall season from 1877 to 1885, will be found recorded in the following table:

Fall of—	Mean Fall Direction.	Velocity for Fall.	Rainfall for Fall.	Clear Days.	Fair Days.	Cloudy Days.	Days Rain Fell.
1877	S.	10.669	1.80	76	6	9	12
1878	N.	11.269	1.35	71	16	4	7
1879	S.	10.492	2.93	59	20	12	12
1880	N.	11.518	.05	71	14	6	2
1881	N.	12.993	2.73	73	15	3	11
1882	N.W.	12.213	6.42	61	22	8	15
1883	S.	10.771	2.48	67	18	6	11
1884	N. & S.E.	10.659	2.61	75	13	3	7
1885	S.E.	14.214	11.44	51	23	17	20
Totals		104.798	31.81	604	147	68	97
Averages	N.	11.644	3.534	67.1	16.3	7.6	10.8

The tabulated statement below shows the number of times the wind was observed blowing from the different points of the compass for the Winter seasons from 1877-8 to 1885-6—three observations daily:

Winter of—	N.	N.E.	E.	S.E.	S.	S.W.	W.	N.W.	Calm.
1877–78	66	7	15	69	54	19	3	11	26
1878–79	102	4	5	41	41	18	4	11	44
1879–80	41	13	19	77	30	17	6	47	23
1880–81	50	5	7	107	57	15	2	19	8
1881–82	86	9	8	75	35	18	7	12	20
1882–83	54	8	16	66	33	14	9	60	9
1883–84	63	11	15	75	37	8	8	40	15
1884–85	42	6	8	68	40	21	9	72	3
1885–86	7	10	4	82	36	23	3	88	17
Totals	511	73	97	660	363	153	51	360	165
Averages	56.8	8.1	10.8	73.3	40.3	17.0	5.7	40.0	18.3

The following table shows the number of times the wind was observed blowing from the different points of the compass; also the number of calms observed during the Spring season, from 1878 to 1885, and is as follows from three daily observations:

Spring of—	N.	N.E.	E.	S.E.	S.	S.W.	W.	N.W.	Calm.
1878	30	2	3	48	89	54	11	23	16
1879	34	3	1	29	82	75	13	16	23
1880	31	4	6	61	59	60	6	45	4
1881	50	4	4	42	71	71	10	14	10
1882	71	0	1	52	56	55	7	21	13
1883	13	3	8	57	91	50	9	42	3
1884	29	2	7	51	70	75	11	27	3
1885	26	3	3	35	66	88	8	40	6
Totals	284	21	33	375	584	528	75	228	78
Averages	35.5	2.6	4.1	46.9	73.0	66.0	9.4	28.5	9.8

The wind's direction for the Summer months from 1878 to 1885 will be found to be as follows from three daily observations:

SUMMER OF—	N.	N.E.	E.	S.E.	S.	S.W.	W.	N.W.	Calm.
1878	13	1	0	25	161	47	6	15	8
1879	24	1	0	5	111	79	8	37	11
1880	14	0	1	46	109	64	10	30	2
1881	21	1	0	58	115	55	6	16	4
1882	3	1	2	56	135	41	9	23	6
1883	6	0	0	54	127	50	9	29	1
1884	2	0	3	52	107	76	8	24	4
1885	5	2	0	66	115	57	6	23	2
Totals	88	6	6	362	980	469	62	197	33
Averages	11.0	0.8	0.8	45.2	122.5	58.6	7.8	24.6	4.8

The table following shows the direction of wind during the Fall months, from 1878 to 1885, both years included, three observations daily:

FALL OF—	N.	N.E.	E.	S.E.	S.	S.W.	W.	N.W.	Calm.
1877	59	10	5	19	78	29	9	31	33
1878	75	5	3	12	64	36	15	37	26
1879	45	9	9	41	66	42	4	28	29
1880	75	7	7	40	55	37	12	30	10
1881	81	10	2	46	46	35	8	28	17
1882	19	8	5	60	54	27	20	66	14
1883	30	4	10	62	66	29	6	53	13
1884	49	7	3	49	47	48	12	38	20
1885	8	8	3	75	58	46	7	65	3
Totals	441	68	47	404	534	329	93	376	165
Averages	49.0	7.6	5.2	44.9	59.4	36.6	10.3	41.8	18.3

The following data shows the number of times the wind was observed blowing from the different points of the compass, and also the number of calms occurring at the time of observations. The calculations are made from three daily observations, making 1,095 observations during each year, and 1,098 for 1880 and 1884, and is as follows:

YEAR OF—	N.	N.E.	E.	S.E.	S.	S.W.	W.	N.W.	Calm.
1878	194	11	12	151	364	155	36	84	88
1879	165	18	14	140	306	220	29	104	99
1880	162	24	33	239	259	172	34	147	28
1881	217	24	12	232	280	176	30	72	52
1882	167	14	21	251	276	145	43	140	38
1883	102	17	37	243	322	138	31	173	32
1884	149	18	22	220	267	213	38	132	39
1885	66	24	12	254	269	210	29	214	17
Totals	1,222	150	163	1,730	2,343	1,429	270	1,066	393
Averages	152.8	18.9	20.4	216.2	292.9	178.6	33.8	133.2	49.1

The following table shows the total number of clear, fair, and cloudy days; also, the number of days in which rain fell for each year, from 1878 to 1885:

Year of—	Clear Days.	Fair Days.	Cloudy Days.	Days on which Rain Fell.
1878	225	81	59	67
1879	223	96	46	69
1880	244	62	60	58
1881	233	68	64	69
1882	251	71	43	62
1883	262	77	26	46
1884	239	69	58	69
1885	228	88	49	57
Totals	1,905	612	405	497
Averages	238.1	76.5	50.6	62.1

WEATHER STATISTICS.

SHOWING THE MONTHLY RAINFALL, ETC., AT SACRAMENTO, FOR 1885, AND JANUARY, FEBRUARY, AND MARCH, 1866.

January, 1885—Mean temperature, normal; rainfall, 2.16 inches, which is 1.61 inches less than the average precipitation for 35 years; frost was deposited on six days; highest water in the river, 23.5 feet; lowest, 17 feet; earthquake shock on the thirtieth.

February—Mean temperature, 4° higher than the average for 33 years; rainfall, .49 of an inch, being 2.44 inches below the average for many years; killing frost twice, and light frost four times; highest river, 18 feet; lowest, 16.1 feet.

March—The mean temperature was about 4° warmer than the average for 33 years. This was the driest month of March known here since 1849. The rainfall was but .08 of an inch, while the average of many years was 3 inches; highest river, 16 feet; lowest, 13.9 feet; partial eclipse of the sun on the sixteenth.

April—The mean temperature was 1.4° above the average for many years; rainfall, .68 of an inch, which was 1.23 inches less than the average for 35 years; two earthquake shocks—one on the third and one on the eleventh; frost on four occasions: highest river, 15.2 feet; lowest, 13.5 feet; last frost of Spring, April twentieth.

May—The mean temperature was 1.5° above the average for 33 years; rainfall, only a sprinkle, which was .74 of an inch below the average precipitation for many years; highest river, 13.9 feet; lowest, 11 feet.

June—The mean temperature was 4.2° below the average for many years; rainfall, very near the average, the latter being .13 of an inch, while the rainfall for this month was .11 of an inch; solar halo on the twenty-fourth; highest river, 11 feet; lowest, 9 feet.

July—This was the windiest month on record, giving from 816 to 2,149 miles more wind than is usual for this month; mean temperature, 2.2° below the average of 33 years; average rainfall for many years, .03 of an inch; this month gave only a sprinkle; two solar halos, and two parhelias, or sun dogs, on the ninth; highest river, 9 feet; lowest, 7.8 feet.

August—The mean temperature was 1.4° above the average for 33 years; the highest temperature recorded occurred this month, and was 105°; rainfall, none, the average of many years being a sprinkle; highest river, 7.8 feet; lowest, 7.3 feet.

September—The mean temperature was 1.2° warmer than the average for 33 years; the average rainfall for many years is .11 of an inch; this month gave .08 of an inch, or a very small amount below the mean; lightning on the fifth and sixth; highest river, 7.10 feet; lowest, 7.3 feet; solar halo on the eighth.

October—The mean temperature was 1.9° warmer than the average for 33 years; the average rainfall for many years was .72 of an inch, while this month only gave .02 of an inch, which is .70 of an inch below the average for many years; sheet and forked lightning on the sixth; solar halo on the twelfth; coronæ around the moon on the twenty-third; highest river, 7.7 feet; lowest, 7.5 feet.

November—The mean temperature was 1.1° greater than the average for 33 years past; there was 2,705 miles more wind than usually sweeps over this station during November; this month was warmer, and gave more wind and a higher velocity, more rain, and more cloudy and rainy days than are usual for the third month of the rainy season. There were 9.44 inches more rain than the average for 36 years; rainfall for the month, 11.34 inches—the average for many years being only 1.90 inches; there were three frosts; highest river, 21.2 feet; lowest, 7.5 feet; first frost of Autumn, November twelfth.

December—The mean temperature was 2.1° warmer than the average of 33 years; the rainfall was 5.76 inches, which is 1.11 inches more than the average precipitation of 36 years. There was a lunar halo on the twenty-fourth, and five light frosts; highest river, 23.9; lowest, 17.2 feet.

We find by the above annual review that the mean temperature for each month was above the average of many years, except June and July; these months were below the average.

The mean temperature for the year 1885, is 61.2°, while the average for 34 years is 60.2°, showing the year 1885 to have been a warmer one than usual. The rainfall from January to June was the least ever known to have been precipitated, and the rainfall from September to December third was the most ever precipitated, except in 1852, when 19.41 inches was measured against 17.21 inches for 1885.

January, 1886—The mean temperature for January was 1.3° below the average of many years; the rainfall was 7.95 inches, which is 4.18 inches in excess of the average precipitation during the last 30 years. There was killing frosts on nine days; and lightning in the northeast on the twentieth, after the heaviest wind storm that has ever visited Sacramento. Lightning, too, in January, is an unheard of thing for this vicinity.

The maximum velocity of wind was over forty-four miles per hour on the twentieth, and on one occasion during the day five miles was made in five minutes, which is at the rate of sixty miles per hour.

The highest river was 25.6 feet on the twenty-eighth; the lowest river was 17.1 feet on the eighteenth. This was the highest water in the river since February 4, 1881, when it reached 26.6 feet. First killing frost of this season was on January second.

February, 1886—The mean temperature for February was 53.3°, which was 2.3° warmer than the average of many years. Average rainfall, 2.866

inches, showing February to have been 2.576 inches below the average of many years. A brilliant meteor passed to the west in the north on the thirteenth; a hailstorm on the twenty-eighth. Highest water in the river, 25 feet, on the first; and the lowest, 14.8 feet, on the twenty-eighth. Peach trees and other fruits in bloom as early as the fifteenth.

March, 1886—Mean temperature, 52.1°, which is 3° cooler than the average mean temperature for 33 years. The rainfall was 2.68 inches, being a trifle less than the average of many years. Heavy hailstorm, with thunder and lightning, on the fourth; lunar halos on the sixteenth and twenty-first; light frosts on the nineteenth and twentieth. Highest river, 19.5 feet, on the sixth and seventh; lowest river, 17.5 feet, on the thirty-first.

WIND.

Gales of twenty-five miles per hour and over, for each month during 1885, and for January, February, and March, 1886, will be found as follows:

January, 1885	None.
February 12, 1885	N.W. 25 miles.
February 23, 1885	N.W. 31 miles.
March, 1885	None.
April, 1885	None.
May 12, 1885	S. 25 miles.
May 22, 1885	N.W. 30 miles.
June, 1885	None.
July 2, 1885	S. 25 miles.
August, 1885	None.
September, 1885	None.
October, 1885	None.
November 16, 1885	S.E. 27 miles.
November 17, 1885	S.E. 36 miles.
November 22, 1885	S.E. 32 miles.
November 24, 1885	S.E. 32 miles.
December 7, 1885	N.W. 25 miles.
January 1, 1886	N.W. 38 miles.
January 20, 1886	S.E. 44 miles.
January 21, 1886	S. 28 miles.
January 22, 1886	S.E. 28 miles.
January 23, 1886	S.E. 30 miles.
February 15, 1886	N.W. 28 miles.
February 26, 1886	S.W. 26 miles.
March 10, 1886	N.W. 28 miles.
March 11, 1886	N.W. 26 miles.
March 14, 1886	N.W. 37 miles.
March 24, 1886	N.W. 30 miles.
March 25, 1886	N.W. 27 miles.
March 27, 1886	N.W. 30 miles.

OAKLAND WEATHER—FOR THE YEAR 1885, AND GENERAL COMPARISON OF WEATHER FROM 1876 TO 1885.

By J. B. Trembley, M.D.

Observations taken at 7 A. M., 2 P. M., and 9 P. M., of each day. Latitude, 37 degrees 48 minutes 20 seconds north; longitude, 122 degrees 15 minutes 20 seconds west of Greenwich; height of barometer above the sea, 24 feet.

BAROMETRICAL PRESSURE.

Table showing the Mean, Highest, and Lowest Monthly Barometer, also the Monthly Range. Barometer not corrected for Elevation or Temperature.

1885.	Mean Monthly Barometer.	Highest Observed Barometer for the Month.	Lowest Observed Barometer for the Month.	Range for the Month.
January	30.08	30.32	29.80	.52
February	30.07	30.37	29.85	.52
March	30.03	30.25	29.80	.45
April	29.93	30.15	29.96	.59
May	29.90	30.05	29.78	.27
June	29.96	30.11	29.78	.33
July	29.95	30.10	29.85	.25
August	29.85	29.97	29.72	.25
September	29.82	29.98	29.77	.21
October	29.93	30.08	29.72	.26
November	29.91	30.20	29.39	.81
December	30.07	30.26	29.62	.64
Means	29.96	30.15	29.75	.42

BAROMETRICAL RECAPITULATION.

Mean barometer for the year _____ 20.96
Maximum barometer for the year, February 3, 9 A. M. _____ 30.37
Minimum barometer for the year, March 17, 7 P. M. _____ 29.39
Highest monthly range for the year _____ .81
Lowest monthly range for the year _____ .21
Yearly range _____ .98

TEMPERATURE.

Table showing the Mean Temperature of the Months, Warmest and Coldest Days; also the Maximum and Minimum Temperatures, the Greatest and Least Daily Variations, Monthly and Mean Daily Range.

1885.	Mean Temperature of the Month	Mean Temperature of Warmest Days	Mean Temperature of Coldest Days	Maximum Temperature	Minimum Temperature	Greatest Daily Variation	Least Daily Variation	Monthly Range of Temperature	Mean Daily Range of Temperature.
January	49.72	56.33	46.00	60	37	20	2	23	9.13
February	54.10	58.33	51.00	70	41	25	2	29	11.86
March	56.93	64.33	52.66	76	43	28	3	33	12.61
April	58.14	65.33	50.66	78	42	29	6	36	12.16
May	58.97	65.33	56.33	80	49	27	7	31	11.61
June	59.70	62.66	55.66	71	52	15	4	19	9.76
July	63.05	68.66	60.66	84	57	26	5	27	12.06
August	60.96	70.00	57.33	85	53	28	4	32	11.29
September	61.88	68.66	57.33	89	51	36	5	38	15.50
October	59.87	64.00	57.33	75	47	28	6	28	12.54
November	56.83	66.33	50.00	72	43	26	1	29	9.16
December	52.42	61.33	46.33	66	41	17	0	25	10.68
Means	57.71	64.27	53.44	75.5	46.33	25.41	3.75	29.16	11.44

RECAPITULATION OF TEMPERATURE.

Mean temperature of the year _____ 57.71
Mean temperature of the warmest day, August 2 _____ 70.00
Mean temperature of the coldest day, January 3 _____ 46.00
Maximum temperature for the year, September 21, 2 P. M. _____ 89.00
Minimum temperature for the year, January 29, 7 A. M. _____ 37.00

Greatest daily variation, September 21 -------------------------------------- 36.00
Least daily variation, December 21 --- 0.00
Greatest monthly range, September -- 38.00
Least monthly range, June --- 19.00
Average daily range for the year --- 11.44
Average monthly range for the year --- 29.16
Yearly range of temperature --- 52.00

SEASONS.

Mean temperature of Winter -- 51.69
Mean temperature of Spring -- 58.08
Mean temperature of Summer --- 61.23
Mean temperature of Autumn --- 59.52
Difference between the coldest and warmest of Spring months ---------------- 2.04
Difference between the coldest and warmest of Summer months --------------- 3.35
Difference between the coldest and warmest of Autumn months --------------- 5.05
Difference between the coldest and warmest of Winter months --------------- 4.38
Difference between the coldest and warmest months of the year ------------- 13.33

RELATIVE HUMIDITY.

Table showing the Relative Humidity, Precipitation, Weather, and Direction from which the Wind Blew, from January 1, 1885, to December 31, 1885, inclusive.

1885.	Mean Relative Humidity.	Rainfall in Inches.	Number of Clear and Fair Days.	Number of Cloudy Days.	Number of Days in which Rain fell.	Number of Foggy Mornings.	Number of Mornings Overcast.	Number of Mornings Frost.	Wind—1,095 Observations.				
									S.W. and W.	N.W. and N.	N.E. and E.	S.E. and S.	Calms
January	86.68	1.92	17	14	7	6	1	8	9	9	16	11	48
February	86.50	.48	21	7	3	4	2	4	29	13	4	7	31
March	85.46	1.07	21	10	5	2	12	1	39	12	3	5	34
April	85.80	3.12	20	10	9	0	5	1	43	7	3	10	27
May	84.20	.10	23	8	1	0	13	0	50	9	2	8	24
June	86.45	.08	20	10	6	0	16	0	67	2	0	12	9
July	84.80	.02	23	8	2	0	18	0	49	9	1	12	22
August	87.20	.00	19	12	0	0	24	0	43	8	1	10	31
September	84.55	.05	· 20	10	2	1	13	0	34	11	2	13	30
October	88.10	.30	21	10	2	4	12	0	32	11	2	6	42
November	90.75	11.11	13	17	17	1	0	3	20	7	10	30	23
December	93.50	4.33	20	11	8	2	2	10	11	14	9	18	41
Means and sums.	86.74	22.58	238	127	62	20	118	27	426	112	53	142	362

RECAPITULATION OF RELATIVE HUMIDITY FOR THE YEAR EIGHTEEN HUNDRED AND EIGHTY-FIVE.

Mean relative humidity for the year -- 86.74
Highest relative humidity during the year ----------------------------------- 100.00
Lowest relative humidity during the year, February 12, 2 P. M. ------------- 41.05
Greatest variation of humidity in 24 hours, September 29 ------------------- 43.08
Least variation of humidity in 24 hours, June 11 --------------------------- .08
Rainfall in inches during the calendar year -------------------------------- 22.58
Rainfall in inches during the agricultural year, 1884-85 ------------------- 17.95
Rainfall in inches since July 1, 1885 (Bay Nursery) ----------------------- 15.81
Number of clear and fair days --- 238
Number of cloudy days -- 127
Number of days in which rain fell --- 62
Number of foggy mornings --- 20
Number of mornings overcast --- 118
Number of mornings that frost was seen ----------------------------------- 27
Wind, direction from southwest and west ---------------------------------- 426
Wind, direction from northwest and north --------------------------------- 112
Wind, direction from northeast and east ----------------------------------- 53
Wind, direction from southeast and south --------------------------------- 142
Calms -- 332

MONTHLY RAINFALL.

As taken in Oakland by Mr. James Hutchison, of the Bay Nursery, for the Consecutive Years Mentioned.

MONTHS.	1873. Quantity	1874. Quantity	1875. Quantity	1876. Quantity	Days	1877. Quantity	Days	1878. Quantity	Days	1879. Quantity	Days
July				.10	2	.18	2		2		2
August					1			.57	3		2
September				.15	5				3		2
October	.60	2.24	.30	4.74	10	.45	4	1.85	2	.70	5
November	.00	9.18	7.83	.25	2	1.62	8	.65	3	2.98	9
December	10.18	.31	4.10			1.75	8	.31	6	5.06	14
	1874.	1875.	1876.	1877.		1878.		1879.		1880.	
January	5.60	6.15	5.28	4.19	9	10.82	16	3.84	11	1.71	7
February	1.80	.30	4.87	1.42	5	11.63	17	5.65	9	2.19	9
March	5.25	1.65	4.55	.96	7	4.30	16	7.96	15	1.70	9
April	1.25		.93	.22	5	1.18	6	1.17	12	8.46	18
May	.75	.10	.45	.33	5	.40	2	1.39	7	1.04	5
June		1.64	.24		5			.16	1		
Amount	26.03	21.67	28.55	12.36	51	32.33	79	23.55	74	23.84	82

MONTHS.	1880. Quantity	Days	1881. Quantity	Days	1882. Quantity	Days	1883. Quantity	Days	1884. Quantity	Days	1885. Quantity	Days
July						2					.02	2
August									.25	1	.00	0
September			.40	1	.42	2	1.00	2	.35	2	.05	2
October	.05	1	.82	7	2.65	9	1.03	7	2.80	4	.30	2
November	.35	2	1.49	5	4.33	7	.90	3	.05	3	11.11	17
December	12.57	18	5.09	10	1.14	9	1.15	6	7.33	13	4.33	8
	1881.		1882.		1883.		1884.		1885.		1886.	
January	10.48	11	2.42	9	1.95	3	3.81	9	1.92	7	6.77	
February	3.95	14	2.05	10	.70	5	5.25	10	.48	3	.30	
March	.88	3	4.20	11	3.33	8	8.59	11	1.07	5		
April	1.40	8	1.51	8	2.20	8	5.79	10	3.12	9		
May	.50	6	.15	3	3.50	11	.55	5	.10	1		
June	1.16	2		2			3.03	10	.08	6		
Amount	31.24	65	18.13	66	20.22	64	31.10	78	17.95	54	* 22.88	

Season of 1885–86, up to March 1, 1886. * Mean, twelve years, 23.89 inches.

The rainfall in California is rather phenomenal, and depending greatly upon latitude and topography. The average difference of annual rainfall in the State, extending from northwest to southeast, is a little over two inches for every degree, being thirty-four inches at Crescent City, and ten inches at San Diego. It is said to increase about one inch for every hundred feet in elevation in ascending the Sierra Nevada.

MONTHLY METEOROLOGICAL SYNOPSIS FOR THE YEAR 1885.

January—The weather was not unusual for the month—rain and light showers were quite prevalent during the first half, and hazy, frosty mornings the last half. The morning of the twenty-sixth was very hazy, and a white, hard frost covered almost everything that was exposed out of doors. At 1:30 A. M. two shocks of an earthquake were felt; the last seemed to be rather a continuation of the first. A loud, rumbling noise or sound attended the peculiar shaking or quaking motion, which apparently came from the northeast, passing towards the southwest. The first shock was quite severe, the rumbling sound, loud and sharp, gradually growing weaker, but rising again in intensity as the second shock was felt, then stopping short, and nothing more was heard of the earthquake's rumbling noise; windows settled in their frames, and all pendent objects swayed to and fro for a few moments. Another shock occurred at 8:56 P. M. The first sensation was an impulsive movement, a jog, ending with a tremor or trembling; a distant rumbling sound was heard during the vibration, which appeared to pass from northeast to southwest. Lunar halo, 9 P. M.; barometer, 30:10; thermometer, 50 degrees; balance of month weather hazy, with slight rains for several days.

February—No unusual phenomena occurred during the month—some light showers and sprinkles of rain; frosty mornings; lunar and solar halos, with a few beautiful luminous sun-risings and sun-settings.

March had her full amount of bluster, by raising a great dust quite a number of times, with high winds and light showers during the latter part of the month. The fifth was very clear and pleasant in the morning; at 2:30 P. M. a gale of wind came up from the northeast, which filled the air with clouds of dust; the thermometer ran up to summer heat, which, with the dryness of the atmosphere, caused the most succulent plants to wilt for several hours; evening very clear and pleasant. An eclipse of the sun, which it was said to have been annular, occurred on the 16th; no opportunity was given to see it on account of a thick, dark overcast that remained all day; during the time of the greatest obscuration of the sun the darkness of night was very perceptible for a time; a chill was felt in the air; the thermometer fell 4 degrees in temperature and gradually rose again when the eclipse passed off. The forecasting of the weather in the public mind—that eclipses are followed by storms within forty-eight hours—in this case was verified, for a fine shower of rain fell on the 18th, which broke the long continued drought for four or five weeks.

April was what might be called a rainy month; warm showers at frequent intervals seemed to be the prevalent characteristic of the precipitation. On the evening of the 8th, at 7:45 o'clock, there occurred one of the most severe showers or storms that has been observed in Oakland for ten years past. The rain fell in torrents, filling the streets and gutters with water; vivid flashes of lightning illuminated the dark horizon every few minutes, and loud peals of thunder crashed here and there, then rolled away in the distance with muttering sounds. For one and a half hours this terrific storm, swayed by different currents of wind, hung over Oakland with all its fury, then passing away and leaving behind on its track brilliant flashes of heat lightning that lighted up the huge cumulous clouds that carried away the relics of the storm. No other meteorological phenomena occurred during the month, unless it was a shower of rain and hail that fell at 5 P. M. on the 19th. Fine growing weather, with plenty of moisture, was the marked feature in this vicinity.

May, June, and July were more than usually marked by many overcasts or high fogs, lasting late in the morning and coming up early in the afternoon. Very dry weather supervened on the April rains, which in many places dried up and destroyed entire fields of grain, and injuring them more or less in every locality in the State.

August, September, and October were counterparts of the three preceding months—overcasts, drought, and disagreeable weather mornings and evenings were at their maximum for ten years. August, no rain. September, one half an inch fell the day following an eclipse of the moon, which occurred on the 23d, at 10 o'clock P. M., invisible on account of clouds and overcast. Thermometer marked the maximum temperature of the year on the 21st, at 2 o'clock P. M., 89°. During October the precipitation was only .30 of an inch; rather unusual, but not without precedent.

November—The extremes appertaining to the weather culminated on the beginning of this month. Solar and lunar halos preceded the rains, which were very frequent, and gave a rainfall unprecedented in amount for November. On the seventeenth a heavy rainfall prevailed, which poured in torrents for a few hours in the early morning. Large portions of the surface of the streets and vacant grounds were covered with water. Sewers and gutters were overflowed; barometer fell to 29.39, lowest observed in Oakland during the time of observation, ten years; showery during the day and night, until about four o'clock P. M. of the eighteenth, when a peculiar phenomenon occurred by the various air currents which came to and near the surface of the earth. Dark, heavy clouds overcast nearly the whole horizon, and near the surface of the earth, bounding the visible horizon all round, a bank of black, dark clouds gave every appearance of a severe storm. The upper currents of the atmosphere were moving in several directions; but at the surface, and coming over the eastern foothills, a high wind blew from the northeast, passing over the city to the southwest a short distance, when it met a counter current which turned it upward in its course, where it arose high in the air, dispelling the dark clouds, then turning backward upon itself towards the northeast, and descending to the tops of the foothills, covering them with a white, fleecy vapor, in striking contrast with the dark rain clouds in the background. This current moved in a circle of a short distance in diameter at a right angle to the surface of the earth, like some great wheel, rising high enough to cut into and through the dark storm-cloud overhead, dashing and breaking up all clouds it came in contact with, into a white fog or foam. There were three atmospherical currents prevailing at the same time—one from the east, one from the south-southwest, and one from the northwest. Clouds were passing in each of these currents, in the direction to which they were moving, but it was very difficult to account for the circular and vertical current that was seen, and the violent phenomenon it produced for the short time it existed. The great atmospherical currents, when they came in contact with it, disappeared for the time being in the vortex of the circular current, and were lost in its rapid motion on its axis. The month was phenomenal in its great amount of rainfall—11.11 inches.

December—Nothing unusual occurred—light frost, rain and growing weather for vegetation. On the thirtieth, at 9:50 o'clock, a shock of an earthquake was felt, lasting several seconds; vibrations from east to west, half gyratory, then lifting. It was attended with a deep rumbling sound. Pendent objects swayed to and fro, and loose windows rattled in their casements. A heavy mist and light sprinkle of rain prevailed at the time; P. M., clear, cool, and very pleasant.

TABLE

Showing the Comparative Annual Meteorology of 1876, 1877, 1878, 1879, 1880, 1881, 1882, 1883, 1884, and 1885.

	1876.	1877.	1878.	1879.	1880.	1881.	1882.	1883.	1884.	1885.
Mean temperature of the year	55.09	56.29	55.28	55.11	53.69	55.62	54.49	54.66	55.85	57.71
Mean temperature of warmest day	74.	76.	69.33	75.33	70.66	70.	69.33	84.66	72.66	70.
Mean temperatnre of coldest day	36.	41.63	37.	33.66	41.	42.	35.	32.33	36.	46.
Maximum temperature for the year	97.	96.	84.	93.	89.	87.	84.	103.	88.	89.
Minimum temperature for the year	30.	30.	27.	27.	29.	31.	30.	25.	28.	27.
Greatest daily variation of temperature	33.	38.	33.	46.	36.	35.	31.	38.	30.	36.
Least daily variation of temperature	2.	1.	2.	------	1.	1.	1.	1.	1.	00.
Greatest monthly range of temperature	49.	47.	46.	46.	48.	40.	42.	50.	45.	38.
Least monthly range of temperature	19.	25.	23.	30.	28.	21.	19.	29.	19.	19.
Average daily range of temperature for year	14.94	14.61	13.65	12.96	14.10	13.40	12.80	12.81	11.64	11.44
Average monthly range of temperature for year	34.92	35.5	32.5	38.	34.91	32.	31.16	37.58	30.	29.16
Yearly range of temperature	67.	66.	57.	66.	60.	56.	54.	65.	60.	52.
Mean relative humidity for year	83.	83.11	84.71	85.29	83.70	83.25	82.57	83.71	85.39	86.74
Highest relative humidity for year	100.	100.	100.	100.	100.	100.	100.	100.	100.	100.
Lowest relative humidity for year	40.	34.40	38.60	39.	27.	29.	28.7	33.9	38.1	41.5
Greatest variation humidity in 24 hours	49.09	51.20	45.06	58.	54.40	37.40	65.7	48.8	41.	43.8
Least variation humidity in 24 hours	.06	.01	.02	.30	.20	.30	.4	.3	.3	.8
Rainfall in inches during the year	21.56	11.09	31.71	28.91	28.07	26.07	18.87	15.76	38.20	22.58
Rainfall in inches in agricultural years, from July 1, 1876, to July 1, 1885	28.53	12.33	32.32	23.55	23.84	31.24	18.03	20.22	31.10	17.95
Number clear and fair days during year	268	301	255	266	258	276	276	266	260	238
Number cloudy days during year	98	64	110	99	108	89	89	99	106	127
Number days in which rain fell	63	58	78	89	53	67	72	53	85	67
Number foggy mornings	23	8	17	19	27	28	15	21	19	20
Number mornings overcast	51	44	64	63	86	52	77	105	77	118
Number mornings frost was seen	35	35	36	46	62	47	50	58	38	27
Wind, direction from S.W. and W.	342	364	311	355	346	402	345	428	382	426
Wind, direction from N.W. and W.	210	150	173	150	136	136	150	119	128	112
Wind, direction from N.E. and N.	34	63	45	50	59	58	53	29	62	53
Wind, direction from S.E. and S.	163	150	164	126	172	138	143	91	151	142
Calms	340	368	402	372	385	331	404	438	375	362
Seasons.										
Mean temperature of Spring	54.46	55.18	55.73	56.15	52.97	56.35	54.12	54.63	55.59	58.08
Mean temperature of Summer	60.40	61.17	59.36	60.07	58.95	60.27	60.06	61.16	61.89	61.23
Mean temperature of Autumn	57.75	57.67	56.92	57.73	55.86	54.78	56.44	54.25	57.07	59.52
Mean temperature of Winter	48.20	50.39	50.12	47.60	45.38	51.10	46.80	46.20	47.38	51.69
Difference between the warmest and coldest months of Spring	4.40	1.49	3.68	.70	9.91	5.12	5.77	5.60	6.16	2.04
Difference between the warmest and coldest months of Summer	1.09	1.10	.35	1.26	1.88	1.55	1.13	2.78	2.60	3.25
Difference between the warmest and coldest months of Autumn	6.13	7.76	5.93	9.14	7.70	8.79	9.68	10.64	3.99	5.05
Difference between the warmest and coldest months of Winter	5.00	6.09	1.28	5.13	2.37	5.34	2.33	5.98	1.56	4.38
Difference between the warmest and coldest months of the year	16.20	12.25	13.06	15.68	15.78	12.38	14.77	19.26	16.38	13.33

FOR TEN YEARS.

Mean difference between the coldest and warmest months for ten years _____ 19.67
Mean temperature for ten years _____ 55.37
Mean barometer for ten years _____ 29.94
Mean relative humidity for ten years _____ 84.24
Mean annual rainfall in inches for ten years _____ 24.28

SALINAS, MONTEREY COUNTY.

Mean monthly temperature from May, 1872, to December, 1885; also, average monthly and average yearly temperature:

YEAR.	Jan.	Feb.	March.	April.	May.	June.	July.	Aug.	Sept.	Oct.	Nov.	Dec.	Yearly Av'rge.
1872....		47.6	48.1		61.3	63.4	62.4	60.6	59.6	56.4	52.2	47.8	*57.96
1873....	51.9	47.6	48.1				60.3	63.5	61.1	57.0	55.2	49.6	†54.92
1874....	49.1	49.6	49.2	55.0	57.7	59.8	62.8	61.7	59.8	58.6	53.5	46.0	55.24
1875....	47.6	47.8	47.8	53.4	54.8	60.3	58.0	59.9	58.8	59.1	53.8	49.1	54.20
1876....	45.4	47.8	49.4	52.3	54.4	58.1	51.9	58.4	59.0	57.1	51.8	46.3	52.66
1877....	52.4	51.7	54.3	52.9	55.6	60.3	64.1	59.6	56.3	53.9	50.9	48.5	55.04
1878....	49.4	49.8	51.9	53.1	55.9	57.4	57.3	58.1	56.5	55.3	50.8	43.0	53.21
1879....	43.5	50.1	53.1	54.2	55.7	57.8	58.0	59.0	58.9	55.2	49.1	44.5	53.26
1880....	42.6	41.9	44.7	51.4	55.0	51.4	58.6	58.1	59.3	54.1	45.7	50.8	51.13
1881....	49.1	52.4	49.9	54.0	55.3	59.4	58.9	58.2	56.4	52.0	47.4	49.0	53.50
1882....	43.2	42.1	50.1	50.5	55.2	58.4	59.1	57.8	57.2	54.7	48.5	48.3	52.09
1883....	42.1	43.9	53.2	58.9	56.5	61.3	59.3	58.5	60.5	55.1	50.6	48.5	62.37
1884....	46.3	40.9	52.7	53.9	58.2	59.2	59.5	58.5	56.9	50.6	55.0	47.6	53.28
1885....	48.7	50.3	55.1	56.5	57.9	56.4	61.4	58.9	58.9	56.2	52.7	52.0	55.42
Totals .	611.3	615.9	659.5	646.1	733.5	763.2	831.6	830.8	819.2	776.3	717.2	671.5	764.28
Mont'ly av'age.	47.03	47.38	50.73	53.84	56.42	58.71	59.40	59.34	58.51	55.45	51.23	47.96	54.59

* Average for eight months.
† Average for nine months.

SALINAS, MONTEREY COUNTY.

Highest temperature at Salinas from May, 1872, to December, 1885:

YEAR.	Jan.	Feb.	Mar.	April.	May.	June.	July.	Aug.	Sept.	Oct.	Nov.	Dec.
1872					84	90	71	79	. 80	87	82	80
1873	76	70	84		90		78	79	76	87	84	66
1874	66	66	70	70	82	79	77	76	88	85	75	73
1875	66	71	78	80	77	78	70	76	77	79	70	68
1876	62	69	73	79	68	79	76	75	82	76	81	76
1877	77	73	70	64	70	88	76	79	87	79	78	69
1878	67	63	74	72	70	71	70	71	76	85	78	80
1879	64	72	80	70	75	78	72	87	80	90	75	71
1880	70	64	65	64	87	76	72	70	71	86	83	70
1881	68	75	78	76	73	74	86	70	78	70	74	72
1882	65	66	76	70	72	71	72	73	74	81	75	76
1883	66	81	84	68	85	94	74	78	96	76	75	. 75
1884	65	76	72	72	80	72	75	76	75	84	80	66
1885	70	75	82	82	76	73	73	76	82	72	72	74
Highest	76	81	84	82	90	94	86	87	96	90	84	80
and Year	1873	1883	*1873	1885	1873	1883	1881	1879	1883	1879	1873	†1872

* Same in 1883.
† Same in 1878.

SALINAS, MONTEREY COUNTY.

Lowest temperature from May, 1872, to December, 1885:

YEAR.	Jan.	Feb.	Mar.	April.	May.	June.	July.	Aug.	Sept.	Oct.	Nov.	Dec.
1872 ----------	----	----	----	----	47	54	56	52	47	36	32	22
1873 ----------	36	30	36	----	----	----	53	52	46	30	35	36
1874 ----------	31	32	33	43	45	50	52	52	48	46	33	23
1875 ----------	28	32	32	33	48	49	52	51	50	39	38	31
1876 ----------	30	30	32	40	44	40	51	52	51	42	35	31
1877 ----------	21	25	37	44	47	52	54	54	50	35	34	34
1878 ----------	30	28	37	41	48	51	51	52	50	42	30	26
1879 ----------	26	29	32	44	45	52	52	53	50	42	30	20
1880 ----------	25	25	32	41	44	49	52	52	45	41	28	38
1881 ----------	32	36	32	45	45	52	52	51	44	31	29	33
1882 ----------	22	28	32	42	44	53	53	52	50	40	30	28
1883 ----------	20	24	44	40	45	53	54	52	50	38	29	32
1884 ----------	30	25	34	44	50	54	53	54	45	37	40	26
1885 ----------	33	32	36	41	50	51	54	57	46	38	30	32
Lowest -------	20	24	32	33	44	40	51	51	44	30	28	20
and Year --------	1883	1883	*1875	1875	†1876	1875	‡1876	§1875	1881	1873	1880	1879

* Same in 1876, 1879, 1880, 1881, and 1882.
† Same in 1880 and 1882.
‡ Same in 1877.
§ Same in 1880.

POWAY, SAN DIEGO COUNTY.

Mean monthly temperature at Poway, San Diego County, from November, 1878, to December, 1885; also, the average monthly and yearly temperatures:

MONTH OF—	1878.	1879.	1880.	1881.	1882.	1883.	1884.	1885.	Monthly Averages.
January ------	--------	47.2	46.3	49.7	45.9	50.1	50.5	49.4	48.44
February -----	--------	52.4	44.3	54.0	46.8	51.4	53.0	52.0	50.56
March -------	--------	55.4	48.1	54.2	52.5	46.8	53.8	57.3	52.59
April -------	--------	56.7	55.2	60.7	54.8	56.6	56.6	60.1	57.24
May ---------	--------	58.4	61.4	64.1	62.0	60.7	61.6	63.3	61.64
June --------	--------	65.2	64.7	66.1	64.5	69.3	65.2	65.2	65.74
July --------	--------	68.8	65.0	71.6	68.5	71.1	69.7	70.8	69.36
August ------	--------	70.6	68.2	70.9	71.6	70.3	66.0	75.8	70.49
September ----	--------	67.5	61.0	69.4	67.9	70.6	65.0	69.3	67.24
October ------	--------	60.7	60.5	61.7	60.6	59.3	59.5	63.4	60.81
November ----	53.8	52.6	53.6	54.4	54.7	55.8	54.6	57.7	55.90
December -----	46.1	51.2	53.4	51.3	53.6	54.7	50.0	53.6	51.74
Totals-------	--------	706.7	681.7	728.1	703.4	716.7	705.5	737.9	711.75
Yearly av'ges.	--------	58.89	56.81	60.68	58.62	59.72	58.79	61.49	59.31

4

POWAY, SAN DIEGO COUNTY.

Highest temperature at Poway, San Diego County, California, from November, 1878, to December, 1885:

MONTH.	1878.	1879.	1880.	1881.	1882.	1883.	1884.	1885.
January		79	78	73	75	86	76	71
February		85	67	85	77	82	82	81
March		78	73	87	80	80	71	87
April		86	79	92	80	86	74	88
May		98	93	90	80	93	79	80
June		110	86	89	84	97	101	90
July		88	86	97	89	93	101	97
August		98	93	97	93	94	104	103
September		100	91	101	96	102	90	103
October		97	88	82	89	81	87	97
November	84	84	85	84	85	82	87	82
December	89	77	84	82	88	80	78	85
Highest		110	93	101	96	102	104	103
and Month		June	*May	Sept.	Sept.	Sept.	Aug.	†Aug.

* Same for August.
†Same for September.

POWAY, SAN DIEGO COUNTY.

Lowest temperature from November, 1878, to December, 1885:

MONTH.	1878.	1879.	1880.	1881.	1882.	1883.	1884.	1885.
January		25	26	32	27	28	31	30
February		29	29	35	33	28	33	32
March		36	32	39	38	45	40	37
April		42	43	53	42	45	43	46
May		49	47	55	50	49	52	52
June		55	53	56	54	58	56	55
July		57	54	56	58	62	59	60
August		54	52	58	60	61	59	63
September		51	51	54	47	56	53	57
October		38	42	42	41	42	41	41
November	25	30	32	31	34	36	33	38
December	21	28	34	21	32	35	30	35
Lowest		25	26	21	27	28	30	30
and Month		Jan.	Jan.	Dec.	Jan.	*Jan.	Dec.	Jan.

* Same for February.

SAN DIEGO, CALIFORNIA.

Highest and lowest temperatures at San Diego, in each year, for thirteen years:

Year.	Highest, Degrees.	Lowest, Degrees.	Year.	Highest, Degrees.	Lowest, Degrees.
1872	87.0	37.0	1879	99.0	32.0
1873	85.0	37.0	1880	84.0	32.0
1874	90.0	39.0	1881	85.5	36.0
1875	88.0	38.0	1882	83.4	34.2
1876	88.0	39.0	1883	101.0	32.4
1877	94.0	40.0	1884	91.5	36.0
1878	100.0	35.4	1885	89.5	38.2

RECORD OF RAINFALL

FROM SAN DIEGO TO SISKIYOU AND FROM SAN FRANCISCO TO PLACERVILLE.

The rainfall at Poway, San Diego County, was furnished by Adams Chapin, voluntary observer of the United States Signal Service:

RAINFALL AT POWAY, SAN DIEGO COUNTY.

Year	January	February	March	April	May	June	July	August	September	October	November	December	Total for Year	For Season of	Total for Season
1878											.02	1.57			
1879	2.88	1.50	none	1.20	.08	.20	none	none	none	.30	2.75	4.72	13.73	1879-80	15.61
1880	1.13	1.54	1.76	3.10	.09	none	.06	.16	none	.74	.30	3.56	12.44	1880-81	10.43
1881	1.16	.60	2.86	1.14	.03	none	none	.04	.03	1.17	.20	.73	7.96	1881-82	13.39
1882	6.40	2.69	1.13	.90	.04	.09	none	.01	.04	.29	.60	.27	12.46	1882-83	8.47
1883	.94	1.76	1.87	1.36	1.34	none	none	none	none	1.59	none	2.40	11.26	1883-84	29.45
1884	1.59	9.40	6.96	4.81	2.26	.44	none	none	none	.24	.38	5.91	31.90	1884-85	10.09
1885	.72	.35	.34	2.05	.63	.07	none	none	none	.06	2.71	.90	7.83	1885-86	*10.78
1886	6.34	.77													
Totals	21.16	18.61	14.92	14.66	4.47	.80	.06	.21	.07	4.39	6.96	20.06	97.67		88.04
Av'ges	2.645	2.326	2.131	2.094	.638	.114	.009	.030	.010	.627	.870	2.507	13.953		14.674

* Up to March 1, 1886.

SAN DIEGO, SAN DIEGO COUNTY.

This table runs from November 1, 1871, to March 1, 1886. The figures are from the annual reports of the Chief Signal Officer. They show the rainfall by calendar years and seasonal years; also, the totals and averages by months:

Year	January	February	March	April	May	June	July	August	September	October	November	December	Total for Year	Season of	Inches
1871											1.19	1.39			----
1872	.99	1.63	.46	.26	.12	none	none	.18	none	none	none	1.41	5.05	1871-72	6.22
1873	.34	4.15	.11	.10	.01	none	none	1.95	none	none	.77	5.46	12.89	1872-73	8.10
1874	3.11	3.73	1.20	.35	.32	none	.12	noue	.04	.53	.88	.55	10.83	1873-74	15.06
1875	2.38	.37	.45	.12	.20	.02	none	.21	.39	none	2.25	.41	6.80	1874-75	5.75
1876	2.47	2.44	1.78	.06	.05	.05	.03	.06	.03	.08	.04	.15	7.24	1875-76	9.99
1877	1.05	.23	1.44	.26	.43	none	none	none	none	.81	.06	3.89	8.17	1876-77	3.71
1878	1.45	4.83	1.41	2.91	.58	.16	none	none	none	.96	none	1.57	13.87	1877-78	16.10
1879	3.54	1.04	.10	.60	sprin	.07	none	none	none	.29	2.77	6.30	14.71	1878-79	7.88
1880	.61	1.50	1.43	1.34	.06	.06	.09	.32	none	.53	.28	4.15	10.37	1879-80	14.77
1881	.52	.45	1.88	1.35	.04	.05	none	.01	.04	.24	.12	.30	5.00	1880-81	9.26
1882	4.53	2.55	1.02	.45	.18	.07	none	none	.01	.41	.39	.13	9.74	1881-82	9.51
1883	1.09	.95	.41	.31	1.14	.08	none	none	none	2.01	.20	1.82	8.01	1882-83	4.92
1884	1.34	9.05	6.23	2.84	2.17	.31	none	none	.07	none	.11	4.83	26.95	1883-84	25.97
1885	.35	.02	.78	1.20	.61	.06	sprin	.13	sprin	.31	1.56	.70	5.72	1884-85	8.16
1886	7.00	1.50												1885-86	*11.07
Totals	30.77	34.44	18.70	12.15	5.91	.93	.24	2.86	.58	6.17	10.62	33.06	145.35		145.40
Av'ges	1.698	2.296	1.335	.868	.422	.066	.017	.204	.041	.455	.708	2.204	10.382		10.386

* Up to March 1, 1886.

SAN BERNARDINO, SAN BERNARDINO COUNTY.

The rainfall at San Bernardino was furnished by Mr. Sidney P. Waite, of the San Bernardino Water Company, and extends from July, 1870, to March 1, 1886, and is as follows:

Year	January	February	March	April	May	June	July	August	September	October	November	December	Total for Year	Season of	Total for Season
1870							none	none	.02	.09	3.11	.89			
1871	6.91	2.21	.19	.34	.11	.07	none	.04	.13	.60	.88	3.91	15.39	1870-71	13.94
1872	none	2.20	.37	.79	.06	none	none	.18	.04	none	1.17	4.40	9.21	1871-72	8.98
1873	6.50	1.25	.51	.84	.21	none	none	1.06	.02	.01	.74	5.73	16.87	1872-73	15.10
1874	5.51	8.76	1.08	.48	.42	none	none	none	.06	1.82	1.88	2.20	23.81	1873-74	23.81
1875	7.20	0.15	0.22	.07	.05	none	none	none	none	none	7.50	.02	15.21	1874-75	13.65
1876	6.55	1.92	3.41	.44	.03	.03	none	none	none	none	.20	.40	12.98	1875-76	19.90
1877	3.50	4.03	.83	.26	.30	none	none	none	none	.86	.50	3.95	14.23	1876-77	9.52
1878	3.33	6.68	2.57	1.71	.66	.07	none	.07	none	.02	.14	4.70	20.00	1877-78	20.33
1879	3.59	1.00	.50	1.20	.24	.03	.11	.02	.01	.94	3.40	6.50	17.54	1878-79	11.34
1880	1.56	1.33	1.45	5.00	.04	none	none	none	none	.14	.67	8.80	18.99	1879-80	20.36
1881	1.40	.36	1.66	.46	.01	none	none	none	none	.80	.27	.50	5.41	1880-81	13.50
1882	*1.11	2.65	3.30	2.91	none	none	none	none	none	.10	.15	.45	9.67	1881-82	11.54
1883	1.60	1.10	2.82	2.95	none	none	.19	none	.53	.85	.09	2.63	12.76	1882-83	9.17
1884	1.63	12.20	9.95	5.68	3.17	.59	none	none	none	none	.11	3.75	37.01	1883-84	37.51
1885	2.79	.11	.28	1.89	1.69	.19	none	none	none	.39	4.36	1.20	12.90	1884-85	10.81
1886	6.44	2.52												1885-86	†14.91
Totals	59.62	48.47	29.14	25.02	6.99	.98	.37	1.30	.83	6.94	25.28	49.63	241.45		239.66
M'thly average	3.726	3.029	1.943	1.668	.466	.065	.023	.081	.052	.434	1.580	3.102	16.097		15.977

LOS ANGELES, LOS ANGELES COUNTY.

The following figures, from February, 1872, to June, 1877, are from the records of Mr. C. Duncommun, of Los Angeles; from July, 1877, to date, are from the Signal Office records:

Year	January	February	March	April	May	June	July	August	September	October	November	December	Total for Year	Season of	Inches
1872		2.25	.43	.97	.10	none	none	.22	none	none	none	4.42	*8.39	1872-73	14.80
1873	2.08	7.19	.05	none	none	none	none	1.06	none	none	.74	5.74	16.86	1873-74	23.72
1874	5.51	9.77	1.09	.45	.42	none	none	none	.06	1.81	1.89	.20	21.20	1874-75	21.67
1875	17.22	.15	.22	.07	.05	none	none	none	none	none	7.57	.82	26.10	1875-76	26.74
1876	6.54	7.92	3.41	.45	.03	none	none	none	none	.40	none	none	18.75	1876-77	5.28
1877	3.48	.01	.83	.26	.30	none	none	none	none	.86	.45	3.93	10.12	1877-78	21.26
1878	3.33	7.68	2.57	1.71	.66	.07	none	none	none	.14		4.70	20.86	1878-79	11.35
1879	3.50	.97	.40	1.19	.24	.03	none	none	none	.93	3.44	6.53	17.41	1879-80	20.34
1880	1.33	1.56	1.45	5.06	.04	none	sprin	sprin	none	.14	.67	8.40	18.65	1880-81	13.13
1881	1.43	.36	1.86	.46	.01	none	none	sprin	sprin	.82	.27	.52	5.53	1881-82	10.40
1882	1.01	2.66	2.60	1.83	.63	sprin	none	none	sprin	.05	1.82	.08	10.74	1882-83	12.11
1883	1.02	3.47	2.87	.15	2.02	.03	sprin	none	none	1.42	none	2.56	14.14	1883-84	38.26
1884	3.15	13.37	12.36	3.58	.39	1.39	.02	.02	sprin	.39	1.07	4.65	40.39	1884-85	9.25
1885	1.05	.01	.01	2.01	.06	sprin	sprin	sprin	.05	.30	5.55	1.65	10.69	1885-86	†16.76
1886	7.80	1.41												1886-87	
Totals	59.14	58.78	30.10	18.19	4.95	1.52	.02	1.30	.11	7.26	23.47	44.20	239.83		245.07
Av'ges	4.224	3.919	2.150	1.299	.354	.109	.001	.093	.008	.519	1.676	3.157	17.129		17.505

* Total for eleven months.
+ Up to March 1, 1886.

SALINAS, MONTEREY COUNTY.

The rainfall of Salinas, Monterey County, was furnished by Dr. E. K. Abbott, and extends from July, 1872, to March 1, 1886, showing the rainfall by months, years, and seasons; also the averages:

Year	January	February	March	April	May	June	July	August	September	October	November	December	Total for Year	Season of	Inches
1872							none	none	.01	.02	.02	6.80		1872-73	13.45
1873	3.40	2.40	.80	none	none	none	none	none	.10	.10	.20	4.25	11.25	1873-74	11.17
1874	3.42	none	2.15	.05	none	none	none	none	none	1.83	1.42	none	9.77	1874-75	8.59
1875	4.50	.15	.69	none	none	none	none	none	none	none	5.17	2.18	12.69	1875-76	21.69
1876	6.16	3.55	4.52	none	.01	none	.10	none	.05	1.04	.05	none	15.48	1876-77	4.64
1877	2.54	.16	.30	.10	.40	none	none	none	none	.12	1.00	2.39	7.01	1877-78	23.82
1878	7.05	8.77	2.57	1.92	none	none	none	none	.05	.60	.20	.35	21.51	1878-79	10.94
1879	2.42	2.81	1.85	1.69	.82	.15	none	none	1.05	1.08	2.28		14.15	1879-80	13.22
1880	1.65	1.16	1.64	3.90	.46	none	none	none	none	none	.57	5.56	14.94	1880-81	14.07
1881	3.32	2.32	1.26	.66	none	.38	none	none	.10	.28	.67	1.24	10.23	1881-82	12.93
1882	1.78	2.31	4.86	1.01	.49	.19	none	none	.38	1.43	.65	1.95	15.05	1882-83	11.79
1883	1.91	.95	2.26	1.28	1.98	none	none	none	.19	1.19	.25	.90	9.91	1883-84	20.43
1884	1.71	4.49	5.09	3.05	.72	2.66	none	.18	.11	1.79	.28	4.46	24.54	1884-85	9.30
1885	1.09	.05	.19	1.21	.12	none	none	none	.02	.08	6.60	1.30	10.66	1885-86	*14.57
1886	5.10	1.47													
Totals	45.05	30.50	28.18	15.77	5.00	3.38	.10	.18	1.01	9.53	18.16	33.66	177.19		176.04
Av'ges	3.218	2.185	2.168	1.213	.385	.260	.007	.013	.072	.681	1.297	2.404	13.630		13.541

* Up to March 1, 1886.

SAN FRANCISCO.

The rainfall from 1849 to 1875 in the following table were taken from the report of the State Agricultural Society for 1874, and was furnished to that society by Thomas Tennant. The rainfall from 1875 to date is compiled from the reports of the Chief Signal Officer:

Year	January	February	March	April	May	June	July	August	September	October	November	December	Total for Year	Season of	Inches
1849							none	none	none	3.14	8.66	6.20			
1850	8.34	1.77	4.53	.46	none	none	none	none	.33	none	.92	1.05	17.40	1849-50	33.10
1851	.72	.54	1.94	1.23	.67	none	none	none	1.03	.21	2.12	7.10	15.56	1850-51	7.40
1852	.58	.14	6.08	.26	.32	none	none	none	none	.80	5.31	13.20	27.29	1851-52	18.44
1853	3.92	1.42	4.86	5.37	.35	none	none	.04	.46	.12	2.28	2.32	21.14	1852-53	35.26
1854	3.88	8.04	3.51	3.12	.02	.08	none	.01	.15	2.41	.34	.81	22.37	1853-54	23.87
1855	3.67	4.77	4.64	5.00	1.88	none	none	none	none	.67	5.76		26.39	1854-55	23.68
1856	9.40	.50	1.60	2.94	.76	.03	.02	none	.07	.45	2.79	3.75	22.31	1855-56	21.66
1857	2.45	8.59	1.62	none	.02	.12	none	.05	none	.93	3.01	4.14	20.93	1856-57	19.88
1858	4.36	1.83	5.55	1.55	.34	.05	.05	.16	none	2.74	.69	6.14	23.46	1857-58	21.81
1859	1.28	6.32	3.02	.27	1.55	none	none	.02	.03	.05	7.28	1.57	21.39	1858-59	22.22
1860	1.64	1.60	3.99	3.14	2.86	.09	.21	none	none	.19	.58	6.16	20.46	1859-60	22.27
1861	2.47	3.72	4.08	.51	1.00	.08	none	none	.02	none	4.10	9.54	25.52	1860-61	19.00
1862	24.36	7.53	2.20	.73	.74	.05	none	none	none	.40	.15	2.35	38.51	1861-62	49.27
1863	3.63	3.19	2.06	1.04	.26	none	none	none	.03	none	2.55	1.80	14.56	1862-63	13.08
1864	1.83	none	1.52	1.57	.78	none	none	.21	.01	.13	6.68	8.91	21.64	1863-64	10.08
1865	5.14	1.34	.74	.94	.63	none	none	none	.24	.26	4.19	.58	14.06	1864-65	24.73
1866	10.88	2.12	3.04	.12	1.46	.04	none	none	.11	none	3.35	15.16	36.28	1865-66	22.93
1867	5.16	7.20	1.58	2.36	none	none	none	none	.04	.20	3.41	10.69	30.64	1866-67	34.92
1868	9.50	6.13	6.30	2.31	.03	.23	none	none	none	.15	1.18	4.34	30.17	1867-68	38.84
1869	6.35	3.90	3.14	2.19	.08	.02	none	none	.12	1.29	1.19	4.31	22.59	1868-69	21.35
1870	3.89	4.78	2.00	1.53	.20	none	none	none	.03	none	.43	3.38	16.24	1869-70	19.31
1871	3.07	3.76	1.29	1.93	.21	none	none	none	.03	.11	3.72	16.74	30.86	1870-71	14.10
1872	4.22	6.97	1.64	1.10	.16	.02	none	none	.14	.21	2.62	7.25	24.33	1871-72	34.71
1873	2.17	4.24	.78	.52	.01	.08	.03	.15	none	.68	1.31	10.12	20.09	1872-73	18.02
1874	4.85	1.83	3.55	1.04	.34	.08	none	none	.83	2.73	5.92	.28	21.45	1873-74	23.98
1875	6.97	.20	1.08	.02	.11	1.01	none	none	none	.24	7.27	4.15	21.05	1874-75	19.15
1876	7.55	4.92	5.49	1.29	.24	.04	.01	.01	.38	3.36	.25	none	23.54	1875-76	31.21
1877	4.32	1.18	1.08	.26	.18	.91	.02	none	none	.65	1.57	2.66	11.93	1876-77	11.04
1878	11.97	12.52	4.56	1.06	.16	.01	.01	none	.55	1.27	.57	.58	33.26	1877-78	35.17
1879	3.52	4.90	8.75	1.89	2.35	.05	.01	.02	sprin	.78	4.03	4.46	30.76	1878-79	24.46
1880	2.23	1.87	2.08	10.06	1.12	none	none	none	none	.05	.33	12.33	30.07	1879-80	26.63
1881	8.69	4.64	.90	2.00	.22	.69	none	none	.25	.54	1.94	3.85	23.72	1880-81	20.86
1882	1.68	2.96	3.45	1.22	.21	.04	none	none	.26	2.66	4.18	2.01	18.67	1881-82	16.14
1883	1.92	1.04	3.01	1.51	3.52	.01	none	none	.42	1.48	1.60	.92	15.43	1882-83	20.12
1884	3.94	6.65	8.24	6.33	.23	2.57	sprin	.04	.33	2.55	.26	7.68	38.82	1883-84	32.42
1885	2.53	.30	1.01	3.17	.04	.19	.06	sprin	.11	.72	11.78	4.99	24.90	1884-85	18.12
1886	7.42	.24												1885-86	†25.26
Totals	190.50	133.65	115.51	70.04	23.08	5.59	.42	7.1	5.97	31.50	109.23	197.25	857.79		858.23
Av'ges	5.149	3.612	3.209	1.946	.641	.155	.011	.019	.161	.851	2.952	5.331	23.827		23.840

† Up to March 1, 1886.

OAKLAND, ALAMEDA COUNTY.

The rainfall record below was taken by Mr. James Hutchison, of the Bay Nursery, Oakland. It shows the rainfall by months, by years, and by seasons, along with the monthly totals and averages, extending from October, 1873, to March 1, 1886:

Year	January	February	March	April	May	June	July	August	September	October	November	December	Total for Year	Season of	Total for Season
1873										.60	.60	10.18		1873-74	26.03
1874	5.60	1.80	5.25	1.25	.75	none	none	none	none	2.34	9.18	.31	26.48	1874-75	21.67
1875	6.15	.30	1.65	none	.10	1.64	none	none	none	.30	7.83	4.10	22.07	1875-76	28.55
1876	5.28	4.87	4.55	.93	.45	.24	.10	none	.15	4.74	.25	none	21.56	1876-77	12.36
1877	4.19	1.42	.96	.22	.33	none	.18	none	none	.45	1.62	1.75	11.12	1877-78	32.33
1878	10.82	11.63	4.30	1.18	.40	none	none	none	.57	1.85	.65	.31	31.71	1878-79	23.55
1879	3.84	5.65	7.96	1.17	1.39	.16	none	none	none	.70	2.98	5.06	28.91	1879-80	23.84
1880	1.71	2.19	1.70	8.46	1.04	none	none	none	none	.05	.35	12.57	28.07	1880-81	31.34
1881	10.48	3.95	.88	1.40	.50	1.16	none	none	.40	.82	1.49	5.09	26.17	1881-82	18.13
1882	2.42	2.05	4.20	1.51	.15	none	none	none	.42	2.65	4.33	1.14	18.87	1882-83	20.22
1883	1.95	.70	3.33	2.20	3.50	none	none	none	1.00	1.03	.90	1.15	15.76	1883-84	31.10
1884	3.81	5.25	8.59	5.79	.55	3.03	none	.25	.35	2.80	.05	7.73	38.20	1884-85	17.72
1885	1.92	.48	1.07	3.12	.10	.08	.02	none	.05	.30	11.11	4.33	22.58	1885-86	*22.86
1886	6.77	.30													
Totals	64.94	40.59	44.44	27.23	9.26	6.31	.30	.25	2.94	18.63	41.34	53.72	291.50		286.84
Av'ges	4.995	3.122	3.703	2.269	.772	.526	.025	.021	.245	1.433	3.180	4.132	24.292		23.903

* Up to March 1, 1886.

SACRAMENTO, SACRAMENTO COUNTY.

The following table of rainfall at Sacramento, from September, 1849, to December 31, 1885, was collated from the records of Dr. T. M. Logan, Dr. F. W. Hatch, and those of the United States Signal Service office:

Year	January	February	March	April	May	June	July	August	September	October	November	December	Total for Year	Season of	Inches
1849									.25	1.50	2.25	12.50		1849-50	36.00
1850	4.50	.50	10.00	4.25	.25	none	none	none	none	none	sprin	sprin	19.50	1850-51	4.71
1851	.65	.35	1.88	1.14	.69	none	none	none	1.00	.18	2.14	7.07	15.10	1851-52	17.98
1852	.58	.12	6.40	.19	.30	none	none	none	sprin	none	6.00	13.41	27.00	1852-53	36.36
1853	3.00	2.00	7.00	3.50	1.45	sprin	sprin	none	sprin	sprin	1.50	1.54	19.99	1853-54	20.06
1854	3.25	8.50	3.25	1.50	.21	.31	none	sprin	sprin	1.01	.65	1.15	19.83	1854-55	18.62
1855	2.67	3.46	4.20	4.32	1.15	.01	none	none	sprin	none	.75	2.00	18.56	1855-56	13.76
1856	4.92	.69	1.40	2.13	1.84	.03	none	none	sprin	.20	.05	2.40	14.26	1856-57	10.46
1857	1.38	4.80	.68	sprin	sprin	.35	none	sprin	none	.66	2.41	2.63	12.91	1857-58	15.00
1858	2.44	2.46	2.88	1.21	.20	.10	.01	sprin	sprin	3.01	.15	4.34	16.80	1858-59	16.03
1859	.96	3.91	1.64	.98	1.04	none	none	none	none	.02	6.48	1.83	16.86	1859-60	22.09
1860	2.31	.93	5.11	2.87	2.49	.02	.63	none	.06	.91	.18	4.28	19.19	1860-61	16.10
1861	2.67	2.92	3.32	.48	.59	.14	.55	none	none	sprin	2.17	8.64	21.38	1861-62	35.56
1862	15.04	4.26	2.80	.82	1.81	.01	none	.01	none	.36	sprin	2.33	27.44	1862-63	11.58
1863	1.73	2.75	2.36	1.69	.36	none	none	none	sprin	none	1.49	1.82	12.20	1863-64	7.87
1864	1.08	.19	1.30	1.08	.74	.09	none	.08	sprin	.12	6.72	7.87	19.27	1864-65	22.51
1865	4.78	.71	.48	1.37	.46	none	sprin	none	.08	.48	2.43	.36	11.15	1865-66	17.93
1866	7.70	2.01	2.02	.48	2.25	.10	.02	none	none	sprin	2.43	9.51	26.52	1866-67	25.30
1867	3.44	7.10	1.01	1.80	.01	none	none	none	.01	none	3.81	12.85	30.03	1867-68	32.79
1868	6.04	3.15	4.35	2.31	.27	sprin	none	none	none	none	.77	2.61	19.50	1868-69	16.64
1869	4.79	3.63	2.94	1.24	.65	.01	none	none	sprin	2.12	.85	1.96	18.19	1869-70	13.57
1870	1.37	3.24	1.64	2.12	.27	sprin	sprin	none	sprin	.02	.58	.97	10.21	1870-71	8.47
1871	2.08	1.92	.69	1.45	.76	sprin	sprin	none	sprin	.21	1.22	10.99	19.32	1871-72	24.05
1872	4.04	4.74	1.94	.61	.28	.02	none	none	sprin	.22	1.93	5.39	19.17	1872-73	14.21
1873	1.23	4.36	.55	.51	none	sprin	.02	sprin	none	.31	1.21	10.01	18.20	1873-74	22.90
1874	5.20	1.86	3.05	.89	.37	sprin	sprin	none	.05	2.26	3.80	.44	17.92	1874-75	17.70
1875	8.70	.55	.80	sprin	sprin	1.10	none	none	none	.44	6.20	5.52	23.31	1875-76	26.53
1876	4.99	3.75	4.15	1.10	.15	none	.21	.02	sprin	3.45	.30	none	18.12	1876-77	8.96
1877	2.77	1.04	.56	.19	.64	.01	sprin	sprin	none	.73	1.07	1.43	8.44	1877-78	24.86
1878	9.26	8.04	3.09	1.07	.17	sprin	none	none	none	.55	.51	.47	23.45	1878-79	17.85
1879	3.18	3.88	4.88	2.66	1.30	.13	sprin	sprin	none	.88	2.05	3.41	22.37	1879-80	26.47
1880	1.64	1.83	1.70	14.20	.76	none	sprin	none	none	.05	11.81		31.99	1880-81	16.57
1881	6.14	5.06	1.37	1.64	sprin	.50	sprin	sprin	.30	.55	1.88	3.27	20.71	1881-82	16.51
1882	1.89	2.40	3.78	1.99	.35	.10	sprin	sprin	.57	2.63	3.22	1.13	18.06	1882-83	18.11
1883	2.23	1.11	3.70	.67	2.85	none	none	none	.90	.97	.61	.44	13.48	1883-84	24.78
1884	3.43	4.46	8.14	4.32	.06	1.45	nono	sprin	.60	2.01	none	10.45	34.92	1884-85	16.58
1885	2.16	.49	.08	.68	sprin	.11	sprin	none	.08	.02	11.34	5.76	20.72	1885-86	*28.12
1886	7.95	.29	2.68												
Totals	142.19	103.46	107.82	67.46	25.72	4.59	1.14	.11	4.21	25.80	79.80	172.19	707.07		705.47
Av'ges	3.843	2.796	2.913	1.874	.714	.128	.032	.003	.114	.697	2.217	4.654	19.641		19.596

* Up to April 1, 1886.

RAINFALL AT FOLSOM, SACRAMENTO COUNTY.

The rainfall data tabulated below is from Folsom, Sacramento County, and was furnished by J. H. Sturgis, special River Observer of the United States Signal Service at that point. The rainfall is from September, 1871, to March 1, 1886:

Year	January	February	March	April	May	June	July	August	September	October	November	December	Total for Year	Season of	Total for Season
1871	----	----	----	----	----	----	----	------	sprin	.55	1.95	13.12	------	1871-72	28.82
1872	5.50	4.72	1.60	.63	.75	sprin	none	sprin	sprin	.25	2.80	6.53	22.78	1872-73	15.70
1873	1.64	4.05	.34	.06	.03	none	.01	sprin	sprin	sprin	1.39	10.51	18.02	1873-74	24.45
1874	5.26	2.63	1.82	2.03	.81	sprin	sprin	none	sprin	1.66	5.19	.13	19.53	1874-75	15.70
1875	6.14	.04	1.24	sprin	.07	1.23	none	none	none	.26	7.12	4.49	20.59	1875-76	30.53
1876	5.89	4.06	6.62	1.56	.24	sprin	.26	.03	none	3.76	.25	none	22.67	1876-77	9.90
1877	3.38	.68	.81	sprin	1.02	sprin	sprin	none	none	.75	.54	1.34	8.52	1877-78	25.00
1878	8.41	8.37	4.23	1.10	.26	none	none	sprin	.12	.43	.62	.56	24.10	1878-79	21.91
1879	4.87	4.94	5.43	3.38	1.44	.12	none	sprin	none	1.31	2.20	3.19	26.78	1879-80	25.09
1880	1.51	2.13	1.40	11.39	2.06	none	sprin	none	uone	sprin	.10	9.85	28.44	1880-81	25.91
1881	6.70	6.07	1.38	1.13	sprin	.68	none	none	none	.40	1.21	1.57	22.59	1881-82	18.28
1882	2.38	3.01	3.82	2.51	.27	.06	sprin	none	none	.68	2.81	3.95	20.23	1882-83	22.32
1883	2.11	.80	5.46	1.10	4.57	none	none	none	1.82	1.41	.81	.92	19.00	1883-84	31.02
1884	3.88	5.92	8.14	5.32	1.16	1.64	none	sprin	.64	2.02	none	9.13	37.85	1884-85	16.60
1885	1.91	.84	.15	1.68	sprin	.21	.02	sprin	.21	sprin	10.91	4.88	20.81	1885-86	*24.50
1886	7.60	.99													
Totals	67.18	49.16	42.44	31.88	12.68	3.94	.29	.03	3.87	16.32	39.40	68.84	311.91		311.23
Av'ges	4.479	3.277	3.031	2.277	.906	.281	.021	.002	.258	1.088	2.627	4.589	22.279		22.231

* Up to March 1, 1886.

PLACERVILLE, EL DORADO COUNTY.

The rainfall record at Placerville, El Dorado County, was furnished by Samuel Hale, Superintendent of the El Dorado Water and Deep Gravel Mining Company, and covers a period of seven years and two months, from 1879 to March 1, 1886. Records were also kept from February, 1874, to February, 1877. The total for those years was, for eleven months in 1874, 33.23 inches; 1875, 44.84 inches; 1876, 39.21 inches; January and February, 1877, gave 11.05 inches:

Year	January	February	March	April	May	June	July	August	September	October	November	December	Total for Year	Season of	Total for Season
1879	----	----	----	----	----	----	----	----	----	3.47	5.28	7.53	------	1879-80	52.60
1880	4.38	5.81	4.66	17.52	3.95	none	none	none	none	.35	.58	16.94	54.19	1880-81	48.04
1881	15.53	7.01	3.38	2.36	sprin	1.89	none	none	1.08	2.80	2.87	7.70	44.62	1881-82	42.46
1882	6.71	5.15	9.30	5.53	1.19	.13	sprin	none	.93	5.72	4.94	1.98	41.58	1882-83	36.56
1883	3.74	2.58	6.88	3.54	6.25	none	sprin	none	1.67	3.38	1.67	2.63	32.34	1883-84	57.39
1884	6.06	11.56	14.46	11.82	1.60	2.51	sprin	.03	.85	2.47	.10	22.65	74.11	1884-85	36.53
1885	4.15	.97	.33	3.32	.27	1.42	none	none	.55	none	15.97	5.22	32.20	1885-86	*35.92
1886	13.03	1.15													
Totals	53.60	34.23	39.01	44.09	13.26	5.95	sprin	.03	5.08	18.19	31.41	64.65	279.04		273.58
Av'ges	7.657	4.890	6.502	7.348	2.210	.992	sprin	.005	.847	2.509	4.480	9.236	46.501		45.597

* Up to March 1, 1886.

GEORGETOWN, EL DORADO COUNTY.

The rainfall at Georgetown, El Dorado County, was furnished by C. M. Fitzgerald, of the California Water and Mining Company, and extends from November, 1872, to March 1, 1886:

Year	January	February	March	April	May	June	July	August	September	October	November	December	Total for Year	Season of	Total for Season
1872												4.30	18.72		
1873	4.08	13.05	3.05	3.11	.12	none	.03	none	none	.61	.55	16.60	41.20	1872-73	46.46
1874	16.66	8.03	13.87	5.80	1.32	.20	none	none	none	3.86	14.60	1.24	65.58	1873-74	63.64
1875	17.87	.04	5.07	.31	2.03	2.06	none	none	uone	1.90	24.12	10.85	64.25	1874-75	47.08
1876	13.09	9.97	14.54	4.78	1.22	none	.77	none	none	11.47	.80	none	56.64	1875-76	81.24
1877	12.44	2.14	7.78	1.74	3.87	.24	none	none	none	1.03	4.30	1.97	35.51	1876-77	40.48
1878	16.21	22.78	10.92	2.99	.99	.12	none	none	.66	2.56	2.66	.48	60.37	1877-78	61.31
1879	11.24	12.41	17.57	9.65	3.39	.34	none	none	none	3.85	6.25	11.73	76.43	1878-79	60.96
1880	5.47	6.00	5.50	25.63	5.97	none	2.28	none	none	.18	.37	22.07	71.79	1879-80	70.40
1881	20.83	12.85	3.84	2.40	.40	2.28	none	none	2.02	4.23	3.30	10.32	62.47	1880-81	65.82
1882	8.59	5.88	10.44	7.11	2.06	.18	none	none	.16	7.75	7.00	3.31	52.48	1881-82	54.13
1883	4.70	3.08	8.73	3.87	7.34	none	none	none	1.60	4.10	1.94	3.50	38.86	1882-83	45.94
1884	7.53	13.80	19.94	15.07	1.52	3.65	none	.01	.80	3.54	.03	33.73	99.62	1883-84	72.66
1885	4.37	.82	.24	3.98	.19	2.28	.03	none	1.16	none	20.77	7.03	40.87	1884-85	50.01
1886	18.32	1.16												1885-86	*48.44
Totals	161.40	112.01	121.49	86.44	30.42	11.35	.83	.01	6.40	45.08	90.99	142.15	766.07		760.13
Av'ges	11.529	8.001	9.345	6.649	2.340	.873	.064	.001	.492	3.468	6.499	10.154	58.928		58.472

* Up to March 1, 1886.

GRASS VALLEY, NEVADA COUNTY.

The rainfall that goes to make up the following table for Nevada County was taken at Grass Valley by Mr. Loutzenheiser. It covers a period of twelve years, from 1873 to 1885, inclusive:

Year	January	February	March	April	May	June	July	August	September	October	November	December	Total for Year	Season of	Total for Season
1873	4.01	12.50	1.39	2.32	2.56	none	none	none	none	.83	2.90	19.01	45.61	1872-73	40.00
1874	13.71	6.93	11.71	3.76	1.05	.10	none	none	none	2.95	15.91	1.08	57.20	1873-74	60.09
1875	15.56	1.39	4.14	.29	1.18	2.28	none	none	none	.97	16.99	7.44	50.24	1874-75	44.78
1876	12.01	10.75	12.47	2.80	1.23	.65	none	none	.06	8.72	.02	none	49.31	1875-76	65.31
1877	10.18	2.44	4.79	1.14	1.40	.74	none	none	none	1.21	3.78	1.74	27.42	1876-77	30.09
1878	15.74	17.76	10.18	2.78	.59	none	none	none	.68	2.09	2.54	.75	53.11	1877-78	53.78
1879	10.72	11.51	18.07	7.08	3.08	.30	none	.08	none	2.79	6.54	8.86	69.03	1878-79	56.82
1880	6.40	4.83	4.07	23.31	6.23	.09	none	none	none	.04	.30	22.69	67.96	1879-80	63.20
1881	19.20	8.50	3.33	1.85	.05	1.50	none	none	1.25	3.71	3.52	8.21	51.12	1880-81	57.46
1882	6.03	6.30	7.96	5.27	1.18	.50	none	none	1.88	7.88	4.78	2.83	44.61	1881-82	43.93
1883	3.05	2.97	9.25	2.38	5.77	none	none	none	1.44	3.03	1.48	2.31	31.68	1882-83	40.70
1884	7.80	10.27	13.98	10.98	1.00	2.30	none	none	.98	3.30	.05	28.39	79.05	1883-84	54.59
1885	3.65	1.76	.83	3.17	.16	.90	none	none	2.65	none	19.27	6.36	38.75	1884-85	43.10
1886	12.40	1.43												1885-86	*42.11
Totals	140.46	99.34	102.17	67.13	25.48	9.36	none	.08	8.94	37.52	78.77	109.67	665.09		653.94
Av'ges	10.033	7.096	7.859	5.164	1.960	.720	none	.006	.688	2.886	6.059	8.436	51.161		50.303

* Up to March 1, 1886.

WEST BUTTE, SUTTER COUNTY.

The report of rainfall at West Butte, Sutter County, was furnished by A. S. Noyes, and covers a period of six years and two months, from November, 1879, to December, 1885, inclusive:

Year	January	February	March	April	May	June	July	August	September	October	November	December	Total for Year	Season of	Total for Season
1879											2.38	2.25		1879-80	13.25
1880	.62	.75	.75	5.88	.62	none	none	none	none	none	none	5.38	14.00	1880-81	12.20
1881	3.69	1.38	.75	1.00	none	none	none	none	.31	1.12	.38	2.00	10.63	1881-82	12.26
1882	1.88	2.31	2.57	1.19	.50	none	none	none	.25	.88	2.62	.25	12.45	1882-83	12.44
1883	.75	.10	3.06	.88	3.56	none	none	none	.62	.81	none	.19	10.06	1883-84	19.80
1884	3.81	2.12	6.50	3.75	.25	1.75	none	none	.57	1.00	none	4.94	24.69	1884-85	12.13
1885	2.00	.50	.37	2.12	.18	.45	none	none	.18	.56	7.45	3.65	17.46	1885-86	*17.29
1886	4.75	.70													
Totals	17.50	7.95	14.00	14.82	5.11	2.20	none	none	1.93	4.37	12.83	18.66	89.29		82.08
Av'ges	2.500	1.136	2.333	2.470	.852	.367	none	none	.322	.728	1.833	2.606	14.882		13.680

* Up to March 1, 1886.

MARYSVILLE, YUBA COUNTY.

The rainfall from Marysville covers a period of three years, and was furnished by J. S. Dallam, Special River Observer for the United States Signal Service at that point:

Year	January	February	March	April	May	June	July	August	September	October	November	December	Total for Year	Season of	Total for Season
1882									.99	2.42	2.84	1.31		1882-83	20.12
1883	1.64	.61	3.72	.98	5.61	none	none	none	.53	1.29	.94	.54	15.86	1883-84	23.47
1884	3.93	3.84	6.04	4.14	.16	2.06	none	none	.48	2.32	.03	7.64	30.64	1884-85	13.84
1885	1.82	.43	.27	.61	.02	.22	none	none	.30	none	9.90	4.87	18.44	1885-86	*21.28
1886	5.73	.48													
Totals	13.12	5.36	10.03	5.73	5.79	2.28	none	none	2.30	6.03	13.71	14.36	64.94		57.43
Av'ges	3.280	1.340	3.343	1.910	1.930	.760	none	none	.575	1.508	3.428	3.765	21.647		19.143

* Up to March 1, 1886.

OROVILLE.

The rainfall for Oroville was furnished by Mr. Hiram Arents, Signal Service Observer at that place, from September, 1884, to January, 1886:

Year	January	February	March	April	May	June	July	August	September	October	November	December	Total for Year	Season of	Total for Season
1884									2.27	2.08	.05	9.33	*13.73	1884-85	19.13
1885	2.10	.73	.25	1.64	.65	.39	sprin	none	.20	sprin	11.27	5.53	22.76	1885-86	†23.49
1886	6.13	.36													
Totals	8.23	1.09	.25	1.64	.65	.39	sprin	none	2.47	2.08	11.32	14.86	22.76		19.13
Av'ges	4.115	.545	.250	1.640	.650	.390	sprin	none	1.235	1.040	5.660	7.430	22.760		19.130

* Total for September, October, November, and December, 1884. † Up to March 1, 1886.

COLUSA, COLUSA COUNTY.

The rainfall, etc., from Colusa, Colusa County, was furnished by J. D. McNary, Special River Observer at that point. The table gives the rainfall by seasons from 1872–73 to March 1, 1886, and by months only from 1881 to date:

Year	January	February	March	April	May	June	July	August	September	October	November	December	Total for Year	For Season of	Total for Season
1872															
1873														1872-73	33.46
1874														1873-74	11.28
1875														1874-75	19.02
1876														1875-76	19.79
1877														1876-77	9.20
1878														1877-78	33.34
1879														1878-79	13.98
1880														1879-80	19.21
1881	3.70	2.27	.60	1.42	.34	none	none	none	1.19	none	.43	2.51	12.46	1880-81	16.06
1882	1.51	2.56	2.50	1.27	.04	.65	none	none	.23	1.19	1.73	.69	12.37	1881-82	22.62
1883	1.07	.37	2.36	.79	3.23	none	none	none	.68	.68	.11	.10	9.39	1882-83	11.66
1884	4.82	2.30	5.70	2.97	.12	2.88	none	none	.59	1.06	none	5.30	25.74	1883-84	29.75
1885	2.04	.58	.35	1.22	none	.55	none	none	.02	.79	7.69	3.98	17.22	1884-85	11.69
1886	4.57	.20												1885-86	*17.25
Totals	17.71	8.28	11.51	7.67	3.73	4.08	none	none	2.71	3.72	9.96	12.58	77.18		251.16
Av'ges	2.952	1.380	2.302	1.534	.746	.816	none	none	5.42	.744	1.992	2.516	15.436		19.320

* Up to March 1, 1886.

PRINCETON, COLUSA COUNTY.

The record of rainfall at Princeton, Colusa County, was furnished by David Bently, voluntary observer of the Signal Service, United States Army, and covers a period of eleven years and two months, from 1875 to March 1, 1886:

Year	January	February	March	April	May	June	July	August	September	October	November	December	Total for Year	Season of	Total for Season
1875	4.30	.15	.30	none	.05	1.75	none	none	none	.75	1.95	1.85	11.10	1875-76	17.18
1876	2.53	4.40	3.50	1.05	.15	.05	.90	.05	.15	4.60	.40	none	17.78	1876-77	10.20
1877	1.65	1.75	.85	none	.20	.30	.30	none	none	.98	1.63	1.48	9.14	1877-78	27.12
1878	10.43	7.64	2.28	1.01	.65	none	none	1.02	.20	.50	.96	.13	24.82	1878-79	10.73
1879	1.83	1.71	2.44	1.61	1.10	.12	none	.13	none	.07	1.91	2.81	13.73	1879-80	13.27
1880	.95	.90	.95	4.93	.75	none	none	none	none	none	.10	6.85	15.43	1880-81	15.54
1881	4.30	1.78	.83	1.15	.10	.43	none	none	.60	.60	.22	2.51	12.52	1881-82	11.09
1882	1.21	2.54	1.53	1.08	.28	.52	none	none	.18	1.71	2.42	.62	12.09	1882-83	12.05
1883	.65	.23	2.35	1.07	2.82	none	none	none	.58	.64	.10	.14	8.58	1883-84	17.78
1884	4.03	2.35	5.06	2.71	.05	2.12	none	none	1.13	1.10	none	6.03	24.58	1884-85	12.19
1885	1.66	.57	.21	.98	.36	.15	none	none	.12	.60	7.21	4.78	16.64	1885-86	*16.79
1886	3.91	.17													
Totals	37.45	24.19	20.30	15.59	6.51	5.44	1.20	1.20	2.96	11.55	16.90	27.20	166.41		147.15
Av'ges	3.121	2.016	1.845	1.417	.592	.495	.109	.109	.269	1.050	1.536	2.473	15.128		14.715

* Up to March 1, 1886.

41

RED BLUFF, TEHAMA COUNTY.

This table is made up from the Signal Service records, and shows the total rainfall for each calendar year from 1878 to December 31, 1885, and the rainfall by seasons from 1877–78 to March 1, 1886; also the totals for each month, with the averages from the opening of the Signal Office on July 1, 1877, to date:

Year	January	February	March	April	May	June	July	August	September	October	November	December	Total for Year	Season of.	Inches
1877____							.05	.03	none	1.35	3.13	3.98			
1878____	20.71	16.66	4.16	2.21	.89	none	none	none	.42	1.56	1.66	.69	48.96	1877-78	53.09
1879____	3.18	3.67	5.39	2.12	2.18	.30	.04	.28	sprin	.48	6.05	9.95	33.64	1878-79	21.49
1880____	2.01	1.66	1.70	7.05	1.04	none	none	none	none	.08	.14	12.85	26.53	1879-80	20.94
1881____	9.40	2.79	.51	1.83	.79	.51	sprin	none	1.67	1.61	.73	5.69	24.93	1880-81	28.90
1882____	2.81	3.94	2.67	2.12	.33	.15	none	none	.49	2.80	5.07	1.44	21.82	1881-82	21.12
1883__:	.87	.39	2.60	1.96	2.96	none	none	none	1.04	2.68	.74	.52	13.76	1882-83	18.58
1884____	3.55	2.21	7.81	4.31	.18	.97	none	none	.36	.90	.04	7.73	28.06	1883-84	24.01
1885____	1.84	1.19	sprin	.62	.64	1.37	.05	none	2.91	.10	17.05	3.90	29.67	1884-85	14.74
1886____	4.80	.18												1885-86	*28.94
Totals__	49.17	32.69	24.84	22.22	9.01	3.30	.14	.31	6.29	11.56	34.61	46.75	227.37	_____	211.87
Av'ges	5.463	3.632	3.105	2.778	1.126	.412	.016	.034	.699	1.284	3.846	5.194	28.421	_____	26.484

* Up to March 1, 1886.

YREKA, SISKIYOU COUNTY.

The rainfall data extending from April, 1872, to December, 1884, was furnished by the late Mr. L. Autenreith, of Yreka. The record for 1885, January and February, 1886, are from Mr. C. H. Pyle, of Yreka:

Year	January	February	March	April	May	June	July	August	September	October	November	December	Total for Year	Season of.	Total for Season
1872____				.24	.44	none	.14	none	.25	1.55	1.43	3.72			11.90
1873____	1.28	1.77	.40	.90	.60	none	none	none	.44	.55	1.17	2.20	9.31	1872-73	11.90
1874____	3.78	1.62	1.49	.74	.34	.44	none	none	none	1.29	2.16	none	11.86	1873-74	12.77
1875____	4.35	.19	1.23	.17	.51	.30	.07	none	none	3.34	5.29	6.07	21.52	1874-75	19.27
1876____	2.00	1.93	2.07	.42	.65	.20	.32	.19	.90	3.05	.43	.26	12.42	1875-76	22.48
1877____	1.20	3.24	1.48	.74	1.56	.65	.18	none	none	.20	3.64	.95	13.84	1876-77	13.69
1878____	6.12	3.91	2.80	.37	.56	none	.35	.40	.45	.25	1.15	.45	16.81	1877-78	19.30
1879____	1.53	1.41	3.96	1.56	1.42	.39	.22	.15	none	.77	2.32	7 23	20.96	1878-79	12.94
1880____	2.43	.61	1.20	2.23	.41	none	.15	none	none	.13	.10	2.42	9.68	1879-80	17.35
1881____	11.78	2.58	.19	1.48	none	1.65	.59	.26	.30	3.24	.68	1.60	23.35	1880-81	20.18
1882____	1.81	1.96	.42	1.20	1.02	none	none	none	.90	1.88	1.89	2.09	13.17	1881-82	12 23
1883____	1.38	.47	.53	1.26	1.76	none	.33	.25	.33	1.35	.66	2.95	11.27	1882-83	12.74
1884____	2.10	1.20	2.44	1.41	1.40	1.78	1.33	.51	.33	none	.79	6.19	29.48	1883-84	17.46
1885____	1.16	2.94	none	1.12	3.65	1.66	.58	none	.49	.29	6.98	2.10	20.97	1884-85	18.42
1886____	4.03	.91												1885-86	*14.80
Totals__	44.95	24.74	18.21	12.84	14.32	7.07	4.26	1.76	4.39	17.89	28.69	38.23	204.64	_____	201.73
Av'ges	3.211	1.767	1.401	.917	1.023	.505	.326	.126	.314	1.278	2.049	2.731	15.742	_____	15.515

* Up to March 1, 1886.

SCOTT VALLEY, SISKIYOU COUNTY.

The rainfall for Scott Valley, Siskiyou County, was tabulated from the observations taken by Mr. Isaac Titcomb, of Scott Valley, near Fort Jones:

Year	January	February	March	April	May	June	July	August	September	October	November	December	Total for Year	Season of -	Total for Season
1859								.50	.87	1.00	4.33	.75		1859-60	20.28
1860	2.59	1.25	4.12	.75	2.00	.50	1.62	.24	.49	2.22	2.00	5.74	23.52	1860-61	20.65
1861	1.12	2.50	2.50	3.00	.54	.30	none	none	none	.51	11.56	10.63	32.66	1861-62	40.96
1862	9.29	3.75	1.32	2.00	1.00	.80	.10	none	.02	.15	.12	1.90	20.45	1862-63	15.72
1863	4.75	1.75	2.45	2.00	.40	1.93	.25	.09	.40	.25	1.85	6.17	22.29	1863-64	15.60
1864	2.07	.43	.82	2.70	.51	.31	none	.03	.04	.31	6.00	12.75	26.57	1864-65	26.77
1865	1.87	2.40	1.30	.32	.05	.75	.35	.02	1.15	1.33	9.79	1.21	20.54	1865-66	35.65
1866	6.59	3.50	9.20	.02	1.72	.62	.50	.47	none	.08	2.51	11.75	36.96	1866-67	28.38
1867	9.12	2.02	.64	1.34	.44	.01	none	.20	.40	.88	1.75	9.68	26.54	1867-68	23.61
1868	3.06	1.50	3.70	1.14	.18	1.06	none	none	.06	.50	.77	2.80	14.77	1868-69	18.29
1869	5.76	1.13	1.32	3.61	1.52	.69	.13	none	1.00	.01	3.04	3.56	21.77	1869-70	19.87
1870	5.00	2.91	1.73	1.37	1.12	.13	none	none	.01	.02	1.00	3.50	16.79	1870-71	13.91
1871	1.86	2.47	1.62	2.27	.55	.26	.35	none	.37	.05	1.62	7.68	19.10	1871-72	22.87
1872	4.18	6.94	1.40	.34	.25	.03	.01	.01	.41	.16	2.67	3.38	19.78	1872-73	13.84
1873	1.33	3.00	1.05	1.50	.27	.03	.03	.05	.37	.94	1.71	4.49	14.77	1873-74	21.79
1874	6.38	1.80	3.65	1.55	.71	.13	.01	.09	none	1.55	4.33	.43	20.63	1874-75	13.09
1875	3.13	.17	1.79	.35	.75	.12	.38	.05	none	4.45	7.31	7.33	25.83	1875-76	31.09
1876	2.26	3.33	3.94	.71	1.19	.18	.34	1.00	1.02	3.75	.54	.01	18.27	1876-77	18.90
1877	1.71	4.23	3.10	1.23	1.48	.71	.12	.02	.01	.45	.67	1.62	15.35	1877-78	23.36
1878	9.72	6.53	3.74	.27	.20	.12	.01	.06	.36	2.81	2.16	1.14	27.12	1878-79	26.42
1879	3.25	3.54	8.39	2.66	1.40	.27	.38	.47	.11	.81	4.64	4.58	30.50	1879-80	33.30
1880	10.62	2.32	2.65	5.39	1.32	.02	.37	.07	none	.18	.32	6.76	30.02	1880-81	31.54
1881	13.95	6.53	.79	1.19	.17	1.04	.54	.04	.76	3.53	2.40	4.60	35.54	1881-82	30.03
1882	4.48	5.69	2.22	2.45	1.29	08	2.49	none	1.44	2.86	2.72	3.75	29.47	1882-83	22.27
1883	2.58	1.51	1.11	3.25	2.65	none	.40	.63	.66	2.41	1.11	4.75	21.06	1883-84	27.63
1884	4.28	3.14	3.45	3.06	1.65	.87	1.62	.01	.60	1.04	.16	8.18	28.06	1884-85	22.03
1885	2.50	3.49	.11	1.98	1.40	1.40	1.16	.01	.83	.53	10.24	3.26	26.91	1885-86	*23.40
1886	7.22	1.32													
Totals	130.67	79.20	68.11	46.45	24.76	12.36	11.16	4.72	11.38	32.78	87.32	132.40	625.27		617.85
Av'ges	4.840	2.933	2.620	1.787	.952	.475	.429	.175	.421	1.214	3.234	4.937	24.049		23.763

* Up to March 1, 1886.

RAINFALL BY SEASONS FROM POWAY TO FORT JONES.

The following table shows the rainfall by seasons—making a brief summary of the rainfall at twenty-two different points in this State, extending from San Diego to Siskiyou, and from San Francisco to Georgetown:

Season of—	Poway	San Diego	Los Angeles	Visalia	Salinas	San Francisco	Oakland	Sacramento	Folsom City	Shingle Springs	Placerville	Georgetown	Grass Valley	Marysville	West Butte	Colusa	Princeton	Red Bluff	Reed's Camp	Weaverville	Yreka	Scott Valley
1849-50						33.10		36.00		39.25												
1850-51						7.40		4.71		17.26												
1851-52						18.44		17.98		32.50												
1852-53						35.36		36.36		47.57												
1853-54						23.87		20.98		30.15												
1854-55						23.68		18.62		19.50												
1855-56						21.66		13.76		18.60												
1856-57						19.88		10.46		26.11												
1857-58						21.81		15.00		19.91												
1858-59						22.22		16.03		31.41												
1859-60						22.27		22.09		28.09												20.38
1860-61						19.00		16.10		26.25												20.65
1861-62						49.27		35.56		77.80												40.96
1862-63						13.08		11.58		19.27												15.72
1863-64						10.08		7.87		24.27												15.60
1864-65						24.73		22.51		34.44												26.77
1865-66						22.93		17.93		36.86												35.65
1866-67						34.99		25.30		50.30												28.38
1867-68						38.84		32.79														23.61
1868-69						21.35		16.64														18.29
1869-70						19.31		13.57														19.87
1870-71						14.10		8.47														13.91
1871-72		6.22				34.71	26.03	24.05	28.82													22.87
1872-73		8.10	14.80		13.46	18.02	21.67	14.21	15.70		33.22	46.46	40.00			33.46				54.81		13.84
1873-74		15.06	22.72		11.17	23.98	28.55	22.30	24.45		54.25	63.64	60.09			11.28				21.46	11.90	21.79
1874-75		5.75	21.67		8.30	19.15	12.36	17.70	15.70			47.08	44.78			18.02				41.20	12.77	13.09
1875-76		9.99	26.74		21.59	31.21	32.33	26.53	30.53			81.24	63.31			19.79	17.18			22.02	10.27	31.09
1876-77		3.71	5.28		4.74	11.04	23.55	8.96	9.90			40.48	30.09			9.20	10.90	53.69		52.23	22.48	18.90
1877-78		16.40	21.26		23.83	35.17	23.84	24.86	25.00		52.60	61.31	53.78			13.34	27.12	21.49		32.32	13.60	23.36
1878-79		7.88	11.35		10.94	24.46	31.34	17.85	21.91		48.04	60.96	56.82			13.98	10.73	29.94		61.02	19.30	26.42
1879-80	15.61	14.77	20.34	10.49	13.22	26.63	29.86	26.47	25.09		42.46	70.40	63.20		13.25	19.21	13.27	28.90	95.46	38.56	17.35	33.30
1880-81	10.43	9.26	13.13	3.95	14.07	29.86	18.13	26.57	25.91		36.56	65.82	57.46		12.20	16.96	15.54	21.12	65.00	38.36	20.18	31.54
1881-82	13.39	9.51	10.40	12.81	12.93	16.14	20.22	16.51	18.28		57.39	54.13	43.93		12.26	22.62	11.09	16.58	55.27	50.00	12.23	30.03
1882-83	8.47	4.92	12.11	11.70	11.79	20.12	20.12	18.11	22.32			45.94	40.70	20.12	12.44	11.66	12.05	24.01	71.73	29.70	12.74	22.27
1883-84	29.45	25.97	38.26	6.73	20.43	32.42	31.10	24.78	31.02		36.53	72.66	54.59	23.47	19.80	29.75	17.78	14.74		31.55	17.46	27.63
1884-85	10.69	8.16	9.25	8.17	9.30	18.12	17.72	16.58	16.60			50.01	45.19	13.84	12.13	11.69	12.19	28.94		38.22	18.42	22.03
1885-86*	10.78	11.07	16.76		14.57	25.26	22.86	†28.12	24.50	28.02	35.92	48.44	42.11	21.28	17.29	17.25	16.79			33.40	14.80	23.10
Totals	88.04	145.40	245.07	53.85	176.04	858.23	286.84	705.47	311.23	579.54	361.05	760.13	653.94	57.63	82.08	251.16	147.15	211.87	288.36	511.45	201.73	617.85
Averages	14.674	10.386	17.505	8.975	13.541	23.840	23.903	19.596	22.231	32.195	45.131	58.472	50.303	19.143	13.680	19.320	14.715	26.484	72.090	39.342	15.515	23.763

*Up to March 1, 1886. †Sacramento up to April 1, 1886.

RAINFALL NEAR SANTA BARBARA.

The climatic conditions of Southern California not being so well known as the same conditions are in the central and northern portions of our State, I take pleasure in appending a letter from Mr. Ellwood Cooper, living at Ellwood, near Santa Barbara, Santa Barbara County, written to General W. B. Hazen, Chief Signal Officer of the Army, Washington City, D. C., as giving accurate data on the varying condition of the weather since 1870. The letter is as follows:

SANTA BARBARA, Cal., December 12, 1885.

Brigadier General Hazen, Washington, D. C.:

DEAR SIR: My last report to your department was partially published in the weather review of April, 1884. [The same will be found appended to this letter.—SERGEANT BARWICK.] That report gave the rainfall from 1870 to and including that of the Winter of 1883–84. The rainfall of 1884 and 1885 was 12.56 inches; 9.12 inches falling from October eighth to December thirty-first, and 3.86 inches falling from January first to May.

From my letter containing the information given in the report, as stated above, I laid down the theory that during the Winters when we had heavy rains before January first, we were likely to have light rains after January first. In support of this I called your attention to the Winters of 1871-2, 1878-9, and 1880-1. I have now to add the rainfall of the past Winter, demonstrating the same condition as the three Winters above alluded to. I also wrote in said communication that during the Spring of 1884 we had a series of warm south winds, which caused the unprecedented rainfall of that season, and that since my sojourn in the country, from 1870 down to that time, that the wind had not blown one single hour steadily from that quarter. In my theories there laid down and (?) the statement that by close observation we could, to a certain extent, foretell the probable rainfall each Winter. I now beg to call your attention to the storm of November last, commencing the fifteenth and ending the night of the twenty-fourth. There is no record of so much rain falling in any year, since records have been kept, in the month of November. A very warm wind blew from the southeast (more southerly than easterly), and part of the time due south, the wind on two different days and nights amounting to a gale; many of my fruit trees were uprooted, some broken square off above the ground. This storm commenced apparently without any preparation. In Los Angeles County, 20 miles from the sea, there were no violent winds. I am therefore convinced that there must have been a strong wind blowing from the Gulf of California some time previous to the commencement of the storm here. * * * * * *

Our usual southeast storms cross the country north of Fort Yuma, giving at San Diego about one third as much rain as at Santa Barbara. The storm of November just passed, the greatest amount of rain was condensed between the first and second ranges of mountains. At the base of the Sierra Madre there were 7 inches; at Newhall, 9 inches; in the Ajai Valley, 15 inches; in the Santa Inez Valley, back of Santa Barbara, 19 inches; and at San Luis Obispo, 22 to 24 inches. On the night of the 17th, 9 inches of rain fell in a few hours at the latter place; in the town of Los Angeles, 6 inches; Santa Barbara, 9 inches; at Ellwood (my home), 10 inches; at the south base of the Santa Inez Range, Glen Annie, there were 14 inches; while at the base, on the north side, there were 19 inches. This warm wind, blowing from the mouth of the Gulf of California, was kept westward of the high range on the peninsula, and carried directly over the first ranges from San Pedro to Point Conception. On reaching the second ranges, was met by the cold northwest trades, condensed, and hence the greatest precipitation in the valleys back from the coast. In the Paso Robles country there was not much rain, probably (from the reports), about 4 inches. We have had up to date, since October 15, 10.37 inches of rain; and, according to my theory, we must not expect very much more after January first. I do not predict, but the fact that every Winter since 1870 that gave us 8 inches or more before January first, gave but little after January first. This very strong probability should put farmers and fruit growers on their guard, and they should lose no time in preparing for such an alternative.

I have the honor to be, your obedient servant,

ELLWOOD COOPER.

Since the receipt of the above, Mr. Cooper has furnished General Hazen the following summary:

The review of the rainfall for 1870 to date establishes thus far one unvarying rule, and that is, that in all our rain season, when there has been more than half our Winter average of rain before January first, we have had less after January first, in the ratio or proportion as the amount before was greater. For example:

SEASON OF—	PRECIPITATION.		
	Before January 1.	After January 1.	Total for Season.
1871–72	8.50 inches.	7.44 inches.	15.88 inches.
1878–79	8.12 inches.	6.38 inches.	14.50 inches.
1880–81	13.50 inches.	3.06 inches.	16.56 inches.
1884–85	9.12 inches.	3.44 inches.	12.56 inches.
1885	13.44 inches.		

While I do not pretend to know or to predict how much more rain we will have before the end of Spring, the above table should warn every farmer and fruit grower of the necessity of preparing their work with the expectation of having but little more. The season thus far for the cultivator is the best we have had in sixteen years, and any failure in crops will be the result of neglect on the part of the farmer.

SANTA BARBARA, California, December 31, 1885.

The following is the report spoken of in Mr. Cooper's letter that he had furnished in April, 1884. The table has been brought down to December 31, 1885:

SEASON OF—	PRECIPITATION.		
	Before January 1.	After January 1.	Total for Season.
1870–71	3.00 inches.	6.00 inches.	9.00 inches.
1871–72	8.50 inches.	7.38 inches.	15.88 inches.
1872–73	4.19 inches.	7.31 inches.	11.50 inches.
1873–74	5.75 inches.	9.75 inches.	15.50 inches.
1874–75	4.25 inches.	16.69 inches.	20.94 inches.
1875–76	6.75 inches.	15.88 inches.	22.63 inches.
1876–77	0.63 inches.	6.50 inches.	7.13 inches.
1877–78	5.75 inches.	27.25 inches.	33.00 inches.
1878–79	8.12 inches.	6.38 inches.	14.50 inches.
1879–80	6.37 inches.	21.94 inches.	28.31 inches.
1880–81	13.50 inches.	3.06 inches.	16.56 inches.
1881–82	3.56 inches.	10.94 inches.	14.50 inches.
1882–83	1.31 inches.	11.88 inches.	13.19 inches.
1883–84	3.81 inches.	29.25 inches.	33.06 inches.
1884–85	9.12 inches.	3.44 inches.	12.56 inches.
1885–86	13.44 inches.		
Totals	84.61 inches.	183.65 inches.	268.26 inches.
Average for fifteen seasons	5.641 inches.	12.243 inches.	17.884 inches.

The above table shows that less than one third of the average Winter precipitation occurs before January first, and more than two thirds after that date. In six of the years covered by the above record the rainfall after March first was two inches; in one year it was one inch; in one year there was no rainfall after February; and in five years the rains extended into April and early May. The annual precipitation for California, as shown by the Smithsonian charts, does not indicate the favorable or unfavorable conditions for the production of crops, for the reason that it gives the annual rainfall from January first to December thirty-first, while the result depends upon the rainfall from October to April; that is, the Autumn, Winter, and Spring rains determine the success or failure of the crops. For example, during the Winter of 1876–7 the rainfall after January first was 6.50 inches, and in November and December of the same year it was 5.75 inches, giving a total of 12.25 inches, or a sufficient amount to insure a fair average crop, while in that year the crops were almost a total failure throughout the State. The rainy season of 1883–4 has differed from any of the preceding years. The rains began at the most favorable season—the last of October and in early December—3.81 inches falling before the close of the year. The people of California were never more apprehensive of an impending drought than during January, 1884. Business interests suffered seriously in consequence of the prevalence of this opinion. Many theories were published indicating that the year 1884 would be one of drought; tables were deduced showing such a probability—decades of dry years coming at certain periods, of which this was to be one. Still there never has been a year in which so much rain has fallen after January first as has been the case in this (1884) year. The precipitation for that part of the rainy season preceding January first, compared with the average corresponding season, shows a deficiency of 1.70 inches, while that of the succeeding months exhibits the unusually large excess of 17.64 inches, and the total amount exceeds the largest precipitation of any previous year of the record.

ELLWOOD COOPER.

SANTA BARBARA, California.

RAINFALL FROM FORT JONES, ON THE NORTH, TO POWAY, ON THE SOUTH, FOR JANUARY, 1886.

The following tabulated statement gives the rainfall for January, 1886, for each place named, together with the total rainfall for this season (1885–6) up to February 1, and the rainfall for last season (1884–5) up to an equal date (February 1, 1885), viz.:

LOCALITIES.	Rainfall for Jan., 1886.	Rainfall for Season to Feb. 1, 1886.	Rainfall for last Season to an equal date.
Scott Valley, Siskiyou County	7.22	22.08	12.48
Yreka, Siskiyou County	4.03	13.89	8.47
Red Bluff, Tehama County	4.80	28.76	10.87
Oroville, Butte County	6.13	23.13	No data.
Marysville, Yuba County	5.73	20.80	12.25
Colusa, Colusa County	4.57	17.05	3.99
Princeton, Colusa County	3.91	16.62	9.92
Grass Valley, Nevada County	12.40	40.68	36.37
Placerville, El Dorado County	13.08	34.82	30.22
Georgetown, El Dorado County	18.32	47.28	42.47
West Butte, Sutter County	4.75	16.59	8.51
Nicolaus, Sutter County	5.32	19.69	No data.
Folsom City, Sacramento County	7.60	23.60	13.70
Sacramento, Sacramento County	7.95	25.15	15.22
Oakland, Alameda County	8.12	23.93	13.10
San Francisco, San Francisco County	7.42	25.02	13.35
Salinas, Monterey County	5.10	13.10	7.73
Los Angeles, Los Angeles County	7.80	15.35	7.16
San Diego, San Diego County	7.00	9.57	5.36
Poway, San Diego County	6.34	10.01	7.25

RAINFALL FOR JANUARY, 1886, AND FOR THE SEASON OF 1885–6, UP TO FEBRUARY FIRST.

SIGNAL SERVICE, UNITED STATES ARMY, DIVISION OF THE PACIFIC, }
SAN FRANCISCO, February 1, 1886. }

Table showing the average January and seasonal rainfall, as obtained from a series of years, with the rainfall for the season and month ending January 31, 1886:

STATIONS.	Average, January.	January, 1886.	Average, Season.	Total Season, 1885–86.
Red Bluff	5.55	4.80	16.62	28.77
Tehama	2.93	4.83	9.52	17.73
Chico	4.10	4.44	12.07	19.42
Orland	4.72	4.00	9.20	16.66
Willows	1.96	3.37	7.08	14.62
Williams	2.72	3.83	7.17	14.22
Marysville	3.27	4.76	9.98	16.94
Dunnigan	3.26	8.37	9.15	24.09
Woodland	3.11	4.83	8.60	17.78
Suisun	4.16	8.10	12.32	23.18
Davisville	3.23	5.29	9.61	17.77
South Vallejo	3.10	6.39	8.69	19.06
Martinez	2.57	5.39	8.30	17.79
Napa	5.44	8.09	12.08	21.61
Calistoga	7.48	9.39	18.78	31.20
Antioch	2.11	4.54	6.27	11.60
Ione	2.86	5.15	8.91	15.77
Oakland	3.84	6.77	10.39	20.36
Niles	2.99	6.15	9.53	17.20
Pleasanton	3.20	4.25	8.48	13.80
Livermore	2.52	4.50	7.57	12.69
Tracy	1.68	3.15	5.15	9.60
Lathrop	2.02	3.41	5.60	10.86
Galt	2.57	5.17	7.37	13.06
Brighton	2.73	5.49	7.42	15.59
Sacramento	3.74	7.95	12.11	25.15
San Francisco	4.83	7.42	14.28	25.08
San Mateo	4.07	6.20	10.54	15.57
Menlo Park	2.42	4.97	7.60	13.47
San José	2.55	3.59	6.44	13.15
Gilroy	4.49	6.09	11.16	15.31
Hollister	2.46	3.93	6.35	10.23
Pajaro	4.03	6.05	10.32	18.22
Monterey	2.15	3.09	7.39	11.55
Salinas	3.01	5.18	7.01	12.89
Soledad	2.14	2.44	4.67	9.88
Santa Cruz	4.77	7.60	13.59	21.00
Modesto	1.70	2.73	5.49	8.63
Turlock	1.15	2.52	4.86	10.25
Merced	2.34	2.64	6.06	9.54
Borden	1.33	3.38	4.27	13.00
Fresno	1.29	2.38	4.42	12.26
Kingsburg	1.07	2.04	4.57	10.80
Goshen	0.98	1.71	3.77	7.43
Lemoore	1.66	3.21	5.58	12.62
Tulare	1.41	1.41	3.30	6.31
Delano	0.79	0.55	2.61	5.85
Sumner	0.95	0.85	2.43	4.10
Caliente	1.34	1.36	4.34	6.67
Keene	1.68	2.14	5.37	6.67
Tehachapi	1.46	1.28	4.31	5.86
Mojave	0.70	3.45	2.54	4.86
Ravenna	1.35	5.28	5.16	11.45
Newhall	1.97	5.22	6.75	16.48
San Fernando	2.02	6.70	6.46	15.81
San Luis Obispo	4.61	5.80	12.14	22.49
Los Angeles	2.07	7.80	7.82	15.35
Anaheim	1.37	4.62	4.61	8.71
Spadra	2.91	1.94	6.33	10.15
Mammoth	0.17	0.57	1.41	2.20
San Diego	1.70	7.00	5.80	9.69

A RAINFALL TABLE FOR CALIFORNIA,

AT ONE HUNDRED AND THIRTY-FIVE DIFFERENT POINTS.

The following is a report of the average rainfall for February for from one to thirty-seven years, and the total rainfall for February, 1886; also, the average precipitation of many seasons, up to and including the last day of February, along with the total rainfall for the present season up to March first. The data shows that considerable more rain has fallen this season than the average amount due for many seasons up to an equal date, with the exception of nine stations in the following table. Those places are Point Reyes, average 13.04 inches, and for this season only 12.40 inches; Pigeon Point, average 10.45 inches, and for the present season 10.35 inches; Mojave, average 4.15 inches, and for this season only 3.61 inches; Keeler, average 2 inches, this season, 1.97 inches; Keene, average 8.55 inches, and for this season 7.19 inches; Bishops Creek, average .58 of an inch, this season .24 of an inch; Point Conception, average 8.10 inches, and for this season 5.26 inches; Colton, average 6.25 inches, and for this season 5.62 inches; Westport, average 38.71 inches, for this season 37.82 inches.

The records from which these tables were compiled are those of the Southern Pacific Railway, voluntary observers, Post Surgeons, and Signal Service Stations, which give a good and comparable record of the rainfall of this State, from latitude 32 to 42, and from the sea level to an altitude of from 7,000 to 8,000 feet.

The tabulated matter was prepared at the Signal Service headquarters of the Pacific Coast, at San Francisco, by Lieut. W. A. Glassford, U. S. A., assistant officer in charge. The stations in this table are numbered according to their geographical position in the State, Crescent City being the extreme northernmost point and San Diego the most southern; the stations or places of observation following each other in their proper order from north to south:

No.	PLACES.	No. of Febs. Computed in Av'ge.	* Average for past Februarys.	February, 1886.	* Average for Season.	Total for Season, 1885–86.
1	Crescent City ----------------	4	11.00	8.19	69.40	90.39
2	Yreka. ----------------	13	1.83	.91	11.89	15.38
3	Fort Jones ----------------	26	3.00	1.32	18.21	24.57
4	Fort Bidwell ----------------	29	2.37	1.22	13.43	18.64
6	Orleans ----------------	----------------	----------------	2.41	----------------	43.53
7	Fort Gaston ----------------	21	8.27	5.29	40.72	57.80
8	Humboldt Lighthouse ----------------	6	4.92	1.97	23.79	31.58
9	Cape Mendocino ----------------	3	2.06	1.37	13.06	19.87

No.	Places.	No. of Febs. Computed in Av'ge.	*Average for past Februarys.	February, 1886.	Average for Season.	Total for Season 1885-86.
23	Colusa	5	1.62	.20	10.36	17.25
24	Williams	8	1.69	.00	8.98	14.22
25	Westport	1	3.79	1.35	38.71	37.82
26	Point Arena	7	3.53	1.19	19.44	34.09
27	Laytonville	2	3.53	.30	29.84	37.12
28	Marysville	15	2.34	.34	12.32	17.28
29	Grass Valley	13	7.53	1.43	35.57	42.11
30	West Butte	6	1.20	.70	9.25	17.29
31	Nicolaus			.49		20.18
32	Knight's Landing	8	2.32	.00	11.29	18.46
33	Emigrant Gap	15	9.20	1.97	33.82	46.85
34	Colfax	15	6.81	.34	30.91	35.04
35	Auburn	15	5.09	.00	23.40	25.57
36	Rockland	15	2.89	.34	14.05	19.49
37	Georgetown	13	8.53	1.16	40.97	48.47
38	Placerville	6	5.51	1.15	30.35	35.92
39	Shingle Springs	19	3.87	.69	24.76	28.02
40	Folsom	14	3.45	.90	16.50	24.52
41	Sacramento	37	2.87	.29	14.30	25 39
42	Galt	8	2.38	.00	10.14	13.93
43	Brighton	8	2.82	.07	10.59	15.66
44	Dunnigan (Yolo County)	9	1.84	.00	11.53	24.09
45	Woodland	9	2.45	.00	11.22	17.43
45½	Davisville	14	2.14	.20	11.89	18.00
46	Suisun	14	2.71	.00	15.30	23.18
47	South Vallejo	9	2.41	.00	11.42	19.06
48	Benicia Barracks	29	2.18	.07	11.24	20.98
49	Calistoga	13	4.75	†	23.67	31.20
50	Napa	9	3.45	.00	15.79	21.61
51	Petaluma	14	3.61	.00	16.92	21.62
52	Point Reyes	7	2.63	.13	13.04	12.40
53	San Rafael	10	5.92	.28	27.77	30.57
54	Ross Valley	2	4.22	.28	30.10	37.00
55	Point Benito	7	3.01	.79	17.31	23.78
56	Angel Island	16	2.08	.12	16.12	24.23
57	Alcatraz Island	20	3.00	.15	13.41	21.30
58	Presidio	31	3.21	.10	14.64	20.50
59	Fort Mason	12	2.06	.00	11.81	21.34
60	San Francisco	36	3.71	.24	18.35	25.32
61	Goat Island	7	2.19	.00	11.70	14.76
62	Farallone Island Lighthouse	5	2.43	.41	12.69	19.92
63	Martinez	8	2.56	.05	11.17	17.84
64	East Brother Island (opposite San Quentin Penitentiary)	7	.95	.00	5.33	8.84
65	Brentwood	6	1.22	.03	7.40	13.17
66	Byron	6	1.38	.00	8.27	12.90
67	Antioch	9	1.40	.00	7.67	11.60
68	Stockton	15	2.13	.05	9.54	11.72
69	Lathrop	8	1.99	.01	7.75	10.99
70	Tracy	15	6.22	.35	6.62	10.99
71	Farmington	8	2.56	.41	10.11	13.33
71½	Ione	8	3.18	.07	12.34	15.84
71⅝	Jackson	8	5.18	.75	20.34	24.53
72	Modesto	15	1.22	.10	6.76	8.54
73	Turlock	7	1.42	.08	6.48	10.33
74	Oakland	8	3.56	.30	14.27	20.66
75	Niles	15	2.76	.63	12.50	17.85
76	Pleasanton	8	3.01	.29	11.60	14.09
77	Livermore	15	2.34	.24	10.01	12.93
78	Point Montara Lighthouse	7	2.72	.26	16.06	20.17
79	San Mateo	12	2.65	.35	13.36	15.92
80	Menlo Park	7	1.71	.37	9.63	13.84
81	Ano Nuevo Lighthouse	7	2.61	.94	14.57	17.42
82	Pigeon Point	7	2.11	.62	10.45	10.35
83	San José	12	1.93	1.12	8.45	14.27
84	Los Gatos			1.34		32.65
85	Gilroy	12	2.99	.32	14.28	15.63
86	Aptos	1	.19	.80	19.27	23.16

No.	PLACES.	No. of Febs. Computed in Av'ge.	*Average for past Februarys.	February, 1886.	*Average for Season.	Total for Season, 1885-86.
87	Santa Cruz	8	4.12	.80	18.03	21.80
88	Pajaro	12	2.85	.47	13.33	18.69
89	Merced	14	1.31	.10	7.39	9.64
90	Borden	10	1.62	.08	6.13	13.68
91	Fresno	8	1.21	.58	5.75	12.84
92	Kingsburg	7	1.35	.24	7.06	11.04
93	Hollister	12	1.81	.22	8.28	10.45
94	Erie			.30		22.60
95	Salinas	12	2.26	1.16	9.44	14.05
96	Monterey	5	2.24	1.14	9.79	12.69
97	Chualar	3	1.04	1.10	7.18	12.50
98	Soledad	12	1.47	.93	6.12	10.81
99	Bishops Creek	2	.00	.00	.58	.24
100	Keeler	2	.62	.10	2.00	1.97
101	Traver			.47		9.08
102	Goshen	8	1.35	.43	5.22	7.89
103	Tulare	11	1.26	.15	4.56	6.51
104	Lemoore	7	1.17	.20	6.96	12.75
105	San Luis Obispo	16	3.81	.81	16.05	23.29
106	Delano	10	1.18	·20	3.79	6.25
107	Sumner	10	.78	.20	3.20	4.30
108	Caliente	10	2.04	.66	6.40	7.56
109	Keene	8	3.14	.64	8.55	7.19
110	Tehachapi	9	2.83	.20	7.15	6.06
111	Mojave	7	1.51	.00	4.15	3.61
112	Santa Maria			.97		13.23
113	Point Conception	7	4.18	1.22	8.10	5.26
114	San Buenaventura	11	3.83	1.04	12.63	16.79
115	Ravenna	6	2.33	.11	8.05	11.58
116	Newhall	9	2.89	.69	9.96	17.17
117	San Fernando	8	3.20	.00	10.18	15.81
118	Cahuenga Valley	3	4.83	1.21	14.16	18.02
119	Los Angeles	7	3.76	1.41	12.17	16.76
120	Spadra	11	2.36	.45	8.93	10.85
121	Anaheim	8	2.59	.82	7.56	9.54
122	Orange			1.83		12.78
123	Santa Monica	1	.02	1.27	14.80	21.52
124	Cucamonga		.91			22.90
125	Colton	9	2.44	.40	6.25	5.62
126	San Gorgonio		1.73			6.73
127	San Bernardino	15	2.87	2.52	11.79	14.91
128	Indio	8	.57		2.18	
129	Mammoth Tank	8	.34	.20	1.97	2.40
130	Yuma	10	.57	.08	2.22	3.71
131	Poway	7	2.55	.77	9.26	10.78
132	San Diego	14	2.42	1.51	8.16	11.20

* February, 1886, is not computed in the average.
† Inappreciable.

SPRING, SUMMER, FALL, AND WINTER.

COMPARATIVE TEMPERATURE TABLES FOR CALIFORNIA, OTHER PORTIONS OF THE UNITED STATES, AND HEALTH RESORTS IN EUROPE AND MEXICO.

The following interesting comparative temperature tables show the mean average temperature for Spring, Summer, Fall, and Winter; also, the average yearly temperature, and the highest and lowest temperature at various celebrated health resorts and other places of note in the United States, Mexico, and Europe. It is quite remarkable that out of 63 stations in various parts of the United States that gives the highest temperature, there are 43 that have a maximum of 100 and over, going to show that there are very few places in the United States but what have at times a very high temperature, well up in the nineties, and very often over 100. In the column of lowest temperature a dash thus (-) before a figure, indicates below zero. It will be seen that 33 out of 63 stations (that show the lowest temperature) have a minimum below zero. . Cities very much farther south than Sacramento or Red Bluff show a very much lower temperature, as witness Santa Fe, which has a temperature as low as 13 below zero; Aiken, South Carolina, down to 3 above; Atlanta, Georgia, 3 above; Chattanooga, 3 above; Knoxville, Tennessee, 14 below zero; Prescott, Arizona, 18 below zero; Jacksonville, Florida, 19; Jerusalem, the Holy City, 25 above. It proves beyond a doubt that California can stand a comparison with places very much farther south than in the Upper Sacramento Valley and central part of the State. The columns of mean temperature for the different seasons of the year are also valuable for ready reference, and proof of the salubrity of the California climate. These tables are drawn from records of many years (and not from one single year), making them very much more valuable as reference tables:

STATIONS.	Average Spring Temperature.	Average Summer Temperature.	Average Fall Temperature.	Average Winter Temperature.	Highest Temperature.	Lowest Temperature.	Average Annual Temperature.
Aiken, South Carolina	61.3	77.4	62.0	45.8	102	3	61.6
Atlanta, Georgia	61.3	76.9	61.9	46.4	98	1	61.8
Atlantic City, New Jersey	47.3	70.4	56.1	33.8	99	-7	51.9
Algiers	66.0	77.0	60.0	55.0	no rec.	no rec.	64.5
Boston, Massachusetts	44.9	69.1	51.1	28.1	101	-13	48.3
Baltimore, Maryland	51.8	74.8	56.9	36.0	101	-6	55.6
Bermuda, Atlantic Ocean	63.7	75.2	71.9	58.8	no rec.	no rec.	67.4
Charleston, South Carolina	64.9	81.2	66.7	50.9	104	13	66.0
Charlotte, North Carolina	59 5	77.7	61.2	43.4	101	-5	60.6
Cadiz, Spain	59.9	70.4	65.3	52.9	no rec.	no rec.	62.1
Cape Henry, Virginia	55.1	76.0	62.2	42.3	103	7	58.9
Cairo, Egypt	73.6	85.1	71.5	58.5	no rec.	no rec.	72.2
Cape May, New Jersey	49.0	71.7	57.6	35.6	91	1	53.6
Chattanooga, Tennessee	60.1	76.6	61.0	44.4	101	3	60.6
Cincinnati, Ohio	54.6	76.1	56.6	36.3	104	-10	55.9
Columbus, Ohio	51.2	73.0	54.4	32.3	103	-20	52.8
Chicago, Illinois	46.0	70.0	51.4	27.8	99	-23	48.8
Cheyenne, Wyoming Territory	40.3	61.8	44.3	23.2	101	-38	44.3
Detroit, Michigan	43.4	68.9	51.4	26.3	100	-24	48.0
Dubuque, Iowa	45.1	69.6	48.7	17.5	101	-31	48.2

Stations.	Average Spring Temperature.	Average Summer Temperature.	Average Fall Temperature.	Average Winter Temperature.	Highest Temperature.	Lowest Temperature.	Average Annual Temperature.
Des Moines, Iowa	46.8	70.3	49.0	17.3	103	-26	48.9
Dodge City, Kansas	52.9	75.3	53.5	30.6	108	-20	53.0
Denver, Colorado	47.6	69.8	49.3	29.9	105	-29	49.2
El Paso, Texas	62.9	80.9	60.7	45.0	113	-5	63.3
Florence, Italy	56.0	74.0	60.7	44.3	no rec.	no rec.	58.8
Funchal, Madeira	64.6	70.9	70.2	62.9	no rec.	no rec.	67.2
Galveston, Texas	69.9	83.5	71.3	55.4	99	18	70.1
Grand Haven, Michigan	40.8	64.7	48.5	24.1	92	-24	47.0
Havana, Cuba	76.2	81.3	78.1	73.0	no rec.	no rec.	77.2
Indianapolis, Indiana	50.1	72.3	54.3	29.0	101	-22	53.2
Jerusalem, Palestine	no rec.	no rec.	no rec.	no rec.	94	25	62.8
Jacksonville, Florida	69.0	81.5	69.8	56.6	104	19	69.3
Key West, Florida	76.8	83.9	78.9	70.8	97	44	77.6
Knoxville, Tennessee	57.3	75.0	57.4	39.7	100	-14	57.3
Louisville, Kentucky	55.7	76.8	57.2	37.3	105	-10	56.8
Lisbon, Portugal	60.0	71.0	62.0	53.0	no rec.	no rec.	61.5
Los Angeles, California	58.4	67.6	62.7	53.5	108	28	60.6
Little Rock, Arkansas	62.6	79.0	62.7	45.9	102	6	62.5
Leavenworth, Kansas	52.2	73.5	53.5	26.3	107	-29	53.4
Mexico City, Mexico	63.4	65.2	60.1	53.6	no rec.	no rec.	60.6
Malta	62.8	78.2	71.0	57.5	no rec.	no rec.	67.4
Mentone, France	60.0	73.0	56.6	49.5	no rec.	no rec.	59.8
Mobile, Alabama	67.2	81.4	67.6	52.6	101	14	67.1
Memphis, Tennessee	61.4	79.5	60.6	42.8	102	2	61.1
Milwaukee, Wisconsin	42.1	67.1	48.0	23.3	98	-25	45.1
Nassau, Bahama Islands	77.7	86.0	80.3	70.7	no rec.	no rec.	78.7
Nice, Italy	56.2	72.3	61.6	47.9	no rec.	no rec.	59.5
New Haven, Connecticut	46.5	70.5	52.8	29.6	100	-4	50.0
New York City, New York	47.6	71.6	54.5	31.5	100	-6	51.2
Norfolk, Virginia	57.0	77.5	60.5	41.9	102	6	59.2
New Orleans, Louisiana	68.9	81.9	69.7	55.9	97	20	69.2
Omaha, Nebraska	49.4	74.2	50.5	24.6	105	-25	49.7
Oakland, California	55.0	60.7	56.4	48.4	103	25	55.1
Pau, France	54.1	70.7	57.4	41.9	no rec.	no rec.	56.0
Pensacola, Florida	67.9	80.6	69.3	56.3	97	15	68.5
Prescott, Arizona	50.7	70.0	52.1	35.5	103	-18	52.3
Philadelphia, Pennsylvania	50.5	73.9	55.3	33.2	102	-5	53.2
Rome, Italy	57.6	72.2	64.0	48.9	no rec.	no rec.	60.7
Red Bluff, California	60.0	80.2	63.5	47.0	110	16	62.7
St. Michaels, Azores	61.2	68.3	62.3	57.8	no rec.	no rec.	62.4
Santa Cruz, Canary Islands	68.9	76.7	74.2	64.6	no rec.	no rec.	71.1
Sandy Hook, New Jersey	47.4	72.0	56.1	32.6	101	-6	52.0
Savannah, Georgia	66.7	81.3	66.8	52.7	105	15	66.9
St. Louis, Missouri	54.8	76.8	55.9	34.1	106	-17	55.4
Springfield, Illinois	52.5	74.2	54.7	32.0	101	-14	53.2
St. Paul, Minnesota	43.8	69.5	46.2	17.4	100	-39	44.0
St. Augustine, Florida	68.7	80.4	71.9	58.2	no rec.	no rec.	69.8
Sante Fe, New Mexico	50.1	70.5	51.3	30.3	97	-13	47.9
Salt Lake City, Utah Territory	49.2	72.6	51.6	31.4	101	-20	51.1
Sacramento, California	59.5	71.7	61.5	48.3	105	21	60.2
San Francisco, California	54.6	58.4	58.2	51.4	95	34	55.7
Salinas, California	53.2	58.6	52.1	50.8	83	28	53.7
Santa Barbara, California	59.4	67.7	63.1	54.3	102	31	61.4
San Diego, California	58.1	66.7	62.7	54.4	101	32	60.5
Visalia, California	59.4	80.8	60.3	45.4	109	18	60.9
Vera Cruz, Mexico	77.0	81.9	78.3	70.9	96	61	77.0
Wilmington, North Carolina	61.9	78.7	63.7	47.8	103	10	63.0
Washington City, D. C.	53.0	75.5	56.5	34.9	104	-14	55.0
Yankton, Dakota Territory	43.3	69.8	46.6	14.5	103	-34	45.7
Yuma, Arizona	70.5	89.9	72.3	56.2	118	22	72.3

THE MODIFYING EFFECTS

THE GREAT DESERTS OF CALIFORNIA AND NEVADA HAVE UPON THE TEMPERATURE OF THE INTERIOR VALLEYS OF THE STATE.

The causes that produce the peculiar climatic conditions in the way of temperature distribution in the great interior valleys of California, are well and ably set forth in an article by the late lamented the Honorable B. B. Redding, and is a portion of a very interesting paper published by the State Agricultural Society in 1877, and is as follows:

In addition to the effects due to latitude, to the Pacific Ocean and its Japan Gulf Stream, the temperature of the State is materially modified by the Colorado, Mohave, and Nevada Deserts, lying south and east of this State. These great reservoirs daily absorbing and daily radiating heat on the south and east, the Gulf Stream giving up its heat on the northwest, together combine to send the isothermal lines nearly as far north as they are in the western part of Europe. Redding, at the northern end of the Sacramento Valley, latitude, 40° 35', longitude, 122° 22', elevation, 558 feet, has a mean annual temperature of 64°, within 2' as warm as Charleston, South Carolina, 8° further south; the latter city having a mean of 66°. Red Bluff, latitude, 40° 10', longitude, 122° 15', elevation, 307 feet, has an annual mean temperature of 66°, the same as Charleston. Chico, in latitude 39° 40', has a mean temperature of 62°, or but 4° less than Charleston.

Coming south through the center of the Sacramento Valley from Redding on the north, to Sumner on the extreme south, the mean annual temperature of the various successive stations show the effect of the radiation of heat in this valley, and the influence of the wind from the cool gulf stream where it flows through the Golden Gate and up the Sacramento River.

The mean annual temperature for the places named will show it very plainly:

Redding has an annual mean temperature of......................................64° 1'
Red Bluff has an annual mean temperature of....................................66° 2'
Chico has an annual mean temperature of..62° 5'
Marysville has an annual mean temperature of...................................63° 6'
Sacramento has an annual mean temperature of...................................60° 5'
Stockton has an annual mean temperature of.....................................62° 0'
Modesto has an annual mean temperature of......................................63° 7'
Merced has an annual mean temperature of.......................................63° 2'
Borden has an annual mean temperature of.......................................66° 4'
Tulare has an annual mean temperature of.......................................64° 1'
Delano has an annual mean temperature of.......................................68° 6'
Sumner has an annual mean temperature of.......................................68° 3'

It will be seen by the mean yearly average that Sacramento is the coolest place in the valley, the temperature increasing both north and south from this point. The breeze from the ocean in the Summer follows up the river and reaches Sacramento each day about 5 P. M., and thus reduces the mean of its temperature. It may be from the same influence that its rainfall is increased above the next stations north and south. The reduction of temperature at Sacramento by the air from the ocean passing through the Golden Gate and up the Sacramento River was noted and commented on by the Rev. J. H. C. Bonté, in his discourse on the northerly winds of the great central valley of California. He said: "These winds are most virulent and desiccating in the extreme north and the extreme south ends of the valley; the atmosphere from the Golden Gate and the bays seem to modify the wind ordinarily in the center of the valley."

The tables of temperature above confirm his inference. Tulare is 4° cooler for the year than the next station south, and 2° cooler than the next station north, which may be referred to the influence of the cold air from the high mountains at whose base it is situated, and to the evaporation from Tulare Lake. Another effect of these deserts is to create a daily sea breeze from the southwest return trade winds that prevail on the coast as surface winds during the Summer months. Each day, after the sun rises over these great deserts, they become heated and increase the temperature of the air over their surface; this air rises, and as the whole current of cool air is from the ocean on the west, it rushes in to fill the vacancy. A gentle southwest wind may be blowing on the coast at night or in the morning; by eleven or twelve o'clock, the full force of the sun's rays is felt in the Nevada Desert—the gentle breeze has increased to a brisk wind, and continues until evening, after the setting sun has withdrawn his rays and the desert has radiated its heat into space. The gentle southwest wind resumes its sway until the next day, when, from the same cause, the high wind is again repeated. Dr. Gibbons, in an article on the climate of San Francisco, says: "Whatever may be the direction of the wind in the forenoon, in the Spring, Summer, and Autumn months it almost invariably works round towards the west in the afternoon. So constant is this phenomenon that in the seven months from April to October, inclusive, there were but three days in which it missed, and those three days were all rainy, with the wind from the south or southwest." He adds: "I cannot discover that in any other spot on the globe the wind blows from one octant one hundred and eighty-six days and from the opposite octant only six days in the year."

The mean Summer temperature at Wadsworth and Brown's Station, on the Nevada Desert, on the line of the Central Pacific Railroad, is 80.3°, and for Brown's, 78.1°. The Summer temperature at Fort Mohave is 92.6°, and at Fort Yuma, 92.7°. The mean temperature of the Mohave Desert for July is 93.1°. Fort Yuma is about five hundred miles south-southeast from Wadsworth. The country intervening is entirely desert.

The indraught of westerly winds from the Pacific in Summer does not appear to be alone sufficient to satisfy the demands of the heat of these great deserts. Lieutenant Wheeler states that on the Mohave Desert "southeast winds are by far the most prevalent in the Summer time." He adds: "It is also easily observed that the clouds and Summer rains come from that direction." From this it would appear that the deserts create an indraught from the Gulf of California, as well as from the Pacific Ocean. I have shown that we are in the latitude of the southwest return trade winds, and that their force is augmented by the effects of the radiation of heat from the deserts on our eastern border. The configuration of the immediate coast near San Francisco, from Point San Pedro to Point Reyes, and the open Golden Gate, cause an increased quantity of this daily sea breeze to pass by and over this city. This increased wind and accompanying fog, coming directly from over the cool Japan Gulf Stream, so lowers the Summer temperature of this city that, as has been shown by Dr. Gibbons and the records of the Smithsonian Institute, there is no other place in the whole territory of the United States, of the same elevation, that has so low a mean temperature; the mean Summer temperature at the Golden Gate being 56°. Another cause affecting the climate of California is in the fact that the Sierra Nevada and Cascade Mountains reach the coast of Alaska, and bend like a great arm around its western and southern shores, thus shutting off or deflecting the polar winds that otherwise would flow down over Oregon and California. The cold winds that reach this State are usually from the northwest, and have had their temperature raised by passing over the Japan Gulf Stream before that gulf stream has been reduced to the temperature we find it while passing our coast. It has been shown that this northwest wind precipitates its moisture by becoming reduced in temperature where it meets the coast of Alaska, British Columbia, and Washington Territory. It passes inland, following the Cascade Mountains where they leave the coast. As it comes south it is heated by coming into warmer latitudes, its capacity to take up moisture is increased, but it finds none in its course to take up. The Cascades, which are a continuation of the Sierra Nevada, direct it into the

SACRAMENTO VALLEY,

Where it meets still greater heat, which the more increases its capacity for containing moisture. It therefore possesses all the desiccating qualities for which it has become famous. Of course its influence as a desiccating wind is only felt in the interior, away from the influence of the ocean. The foregoing are some of the principal causes that give to this portion of the Pacific Coast its peculiar climate. The causes of variation in rainfall, temperature, and course of the wind in localities, can be ascertained by a series of local and general observations, lasting for a sufficient period to warrant conclusions from the mean obtained. Such observations, if taken and condensed, would be of value to the farmer, to the merchant, and, in fact, add to the prosperity of every inhabitant of the State.

CLIMATE OF THE SACRAMENTO AND SAN JOAQUIN VALLEYS AND THE FOOTHILLS.

The climate of the Sacramento Valley and foothills being of great interest just at present and since the holding of the Citrus Fair, January 11, 1886, I thought it a very appropriate time to reprint a portion of an article by the late lamented the Honorable B. B. Redding, published in the State Agricultural Society's Report for 1878. The subject spoken of above is on the general climatic condition of the Sacramento Valley and foothills, from Redding on the north to Sumner on the south, and is as follows:

From Redding, in the northern end, to Sumner, at its southern extremity, is a distance of three hundred and fifty miles. The mean annual average temperature of Redding is 64°. The lowest point to which the thermometer has fallen since a record has been kept was. 27°, in December, 1876. [In all probability it fell lower than that since the above article was written, for in 1883 it fell to 19° at Red Bluff.—Sergeant Barwick.] Sumner, at the southern end of the valley, has an annual average temperature of 68°, and an average rainfall of four inches. The lowest point to which the thermometer has fallen at this place was also 27°, on the same day, in December, 1876. [In December, 1883 (since the above was writ-

ten), the temperature fell to 25°, according to the railroad weather reports.
—SERGEANT BARWICK.] There is a remarkable uniformity in the climate
throughout the Sacramento Valley. In it, a difference of 5° of latitude,
between 35° 30′ and 40° 30′, only lowers the annual average temperature
4.15°. The difference of the annual average temperature between corresponding degrees of latitude in the Atlantic States, at an equal distance
from the ocean, is more than 8°. It has been found that the foothills of
the Sierra, up to a height of about twenty-five hundred feet, have apparently
the same temperature as places in the valley having the same latitude. It
has also been found that with increased elevation there is an increase of
rainfall over those places in the valley having the same latitude, as, for
illustration, Sacramento, with an elevation above the sea of 30 feet, has an
annual average temperature of 60.48°, and an average fall of rain of
between eighteen and nineteen inches, while Colfax, with an elevation of
two thousand four hundred and twenty-one feet, has an annual average
temperature of 60.50°, and an average annual rainfall of from forty-two to
forty-three inches. This uniformity of temperature and increase of rainfall appears to be the law throughout the whole extent of the foothills of
the Sierra, with this variation as relates to temperature, viz.: as latitude is
decreased the temperature of the valley is continued to a proportionally
greater elevation. To illustrate, approximately: if the temperature of
Redding, at the northern end of the valley, is continued up the foothills to
a height of two thousand feet, then the temperature of Sacramento, in the
center of the valley, would be continued up to two thousand five hundred
feet, and that of Sumner, in the extreme southern end of the valley, up to
three thousand feet. The increase of rainfall on the foothills in the latitude of Sacramento, due to elevation, is about one inch to each one hundred feet. South from Sacramento the proportion decreases until, at
Sumner, the increase due to elevation is but half an inch to each one hundred feet. This is shown by the record kept at Fort Tejon, in the Tehachapi
Mountains near Sumner, at an elevation of three thousand two hundred
and forty feet, where the annual rainfall is between nineteen and twenty
inches. There is no record kept at any point in the hills above Redding,
but probably in this latitude the increase due to elevation is about one and
a half (1½) inches to each hundred feet. The increase of precipitation on
the hills at the northern end of the valley gives greater density to the forests, and permits them to grow at lower elevations than in the southern
end of the valley. At the same time the difference in temperature is so
small that the character of the vegetation of the hills at each end of the
valley is not dissimilar. The trees found in the vicinity of Redding, at
the northern end of the valley, below an elevation of five hundred feet, are
not found at the southern end until we pass Caliente, at an elevation of
one thousand three hundred feet. It would seem that the temperature of
the valley prevails up the Sierra to an elevation that equals the height of the

COAST RANGE OF MOUNTAINS.

If a line were drawn parallel to the surface of the ocean from the top of
the Coast Range, east, until it met the flanks of the Sierra, it would mark a
level on the Sierra below which the temperature would not materially differ
from that in the

SACRAMENTO VALLEY.

This fact is probably to be ascribed to the prevailing southwest return
trade winds which blow over the State from the ocean for more than three

hundred days in the year. Passing the summits of the Coast Range, but small portions descend into the valley; the remainder reach the sides of the Sierra at about the level of the summits they have passed.

ARBOREAL VEGETATION.

At the northern end of the valley, at an elevation of five hundred feet above the sea, the most of the California oaks are found; of pines, only the nut or digger pine; the buckeye and chemisal. This is the characteristic arboreal vegetation throughout all these three hundred and fifty miles. Its presence everywhere shows increased rainfall over the valley, and similarity of temperature to that of the valley. Our pasture oak is found at lower elevations in the valley, but always on moist land or near river courses, proving that it demands, in addition to temperature, the increased moisture. In the southern end of the valley this vegetation prevails at higher elevations, because it there finds the proper temperature and moisture. Wherever, on the foothills, any of the trees named constitute the preponderant arboreal vegetation, it is an evidence that the temperature is the same as that of the valley, and plants that can be successfully grown in the valley can be grown to as high an elevation on the hills as these trees abound. If one tree were to be taken as the evidence of this uniformity of temperature, it would be the Sabin's (the nut or Digger) pine. It is never seen in the valley or on the hills below an elevation of about four hundred feet. It is not found at a higher elevation than that in which the temperature is the same as that of the valley. It is never found in groves, but singly, among other trees, yet it prevails throughout these three hundred and fifty miles of foothills. While the vegetation is more dense on the hills at the northern end of the valley, due to increased precipitation, there are also local differences—where there is similarity of soil—due to exposure. Throughout all the lower hills, the greatest number of trees is found on gently sloping eastern, northeastern, and northern hillsides, which necessarily are more moist and cool. The southern aspects contain less trees, because exposed to the direct rays of the sun and to the full force of the prevailing winds.

CROPS SUITABLE FOR CULTIVATION.

Every agricultural product that can be grown in the valleys, including the semi-tropical fruits, can be grown with equal facility in these foothills. Ordinarily the land has to be cleared of the trees found upon it, and cultivation must be continuous, for on the whole western face of the Sierra the native trees, when cut or burned down, are rapidly replaced by a new growth of the same kind. These lands are found to have all of the requisites for the successful growth of orchards. Fruit trees thrive better upon them than on the lands of the valley. None of the many theories advanced as to the cause of the treeless condition of many plains and prairies having ample rainfall, seem to be entirely satisfactory, but experience has demonstrated that orchards grow best and thrive with less artificial aid on lands that in a natural condition are covered with trees. The increasing exports of small fruits, such as strawberries, blackberries, and raspberries, from the vicinity of Newcastle and Auburn, and their superior size and quality, prove that this region is better adapted to their culture than any place yet found on the level lands of the valley. The peaches of Coloma have a State reputation for flavor and size. The apples of Nevada and Georgetown, are equal in size, taste, and keeping qualities to the best imported from Oregon.

OROVILLE ORANGES.

The Oroville oranges have been pronounced equal to the best from Los Angeles.

The vine grows with luxuriance, and bears abundantly wherever it has been planted throughout all this region.

"The wines of Coloma have more than a local reputation. Persons competent to judge assert that wine from grapes grown on the foothills is free from the earthy taste that characterizes much of the wine of the flat land of the valleys. They also express the belief that if ever wine is to be made in California as light as that from the Rhine, and as free from alcohol, the grapes will be grown in the higher elevation of the foothills, where snow falls and remains on the ground a few weeks each season. It is said that the long Summers and great heat of the valleys develop the saccharine matter in the grape, which by fermentation is converted into alcohol."

I will annex to this excellent article of Mr. Redding's on the climate and capabilities of the Sacramento Valley and foothills, a table of mean temperatures, maximum and minimum temperatures, clear, fair, and cloudy days, average cloudiness, and rainfall for the four seasons of the year, taken from the annual report of the Chief Signal Officer for the year 1884. These seasonal means, etc., are calculated from observations of from seven to ten years, and give a better average than if they were only for one year. The headings of the tables will explain themselves and are as follows:

SAN DIEGO.

Mean temperature, maximum and minimum temperature, clear, fair, and cloudy days, average cloudiness, in tenths, and rainfall, by seasons:

SAN DIEGO.	Spring.	Summer.	Autumn.	Winter.	Average Highest, Lowest, and Annual.
Mean temperature	58.1	66.7	62.7	54.4	Annual average temperature, 60.5.
Maximum temperature	99.0	94.0	101.0	82.6	Highest temperature, 101.0.
Minimum temperature	38.0	51.0	38.0	32.0	Lowest temperature, 32.0.
Clear days	26.9	24.3	38.0	33.5	Annual number clear days, 122.7.
Fair days	36.7	48.2	36.5	33.7	Annual number fair days, 155.1.
Cloudy days	28.4	19.5	16.5	23.0	Annual number cloudy days, 87.4.
Average cloudiness, in tenths	4.8	4.6	3.7	4.1	Annual average cloudiness, in tenths, 4.3.
Rainfall	1.91	.30	1.24	6.06	Annual average rainfall, 9.51.

LOS ANGELES.

Mean temperature, maximum and minimum temperature, clear, fair, and cloudy days, average cloudiness, in tenths, and rainfall, by seasons:

LOS ANGELES.	Spring.	Summer.	Autumn.	Winter.	Averages, Highest, Lowest, and Annual Totals.
Mean temperature	58.4	67.6	62.7	53.5	.... Annual average temperature, 60.6.
Maximum temperature	100.0	106.0	108.0	88.2	..Highest temperature, 108.0.
Minimum temperature	35.3	47.0	34.2	28.0	....Lowest temperature, 28.0.
Clear days	36.2	34.9	52.3	47.9	.... Annual number clear days, 171.3.
Fair days	35.1	50.6	32.5	26.6	.... Annual number fair days, 144.8.
Cloudy days	20.7	6.5	6.2	15.7	.... Annual number cloudy days, 49.1.
Average cloudiness, in tenths	4.3	3.4	2.5	3.4	..............Annual average cloudiness, in tenths, 3.4.
Rainfall	4.28	.02	1.57	8.86	Annual average rainfall, 14.73.

SAN FRANCISCO.

Mean temperature, maximum and minimum temperature, clear, fair, and cloudy days, average cloudiness, in tenths, and rainfall, by seasons:

SAN FRANCISCO.	Spring.	Summer.	Autumn.	Winter.	Annual Averages, Highest, Lowest, and Annual Total.
Mean temperature	54.6	58.4	58.2	51.4	.... Annual average temperature, 55.6.
Maximum temperature	86.0	95.2	92.0	70.5	... Highest temperature, 95.2.
Minimum temperature	39.0	48.0	41.0	34.0	....Lowest temperature, 34.0.
Clear days	39.2	23.7	30.0	34.7	.... Annual number clear days, 127.6.
Fair days	32.9	46.8	34.9	29.5	.... Annual number fair days, 144.1.
Cloudy days	19.9	21.5	26.1	26.0	.... Annual number cloudy days, 93.5.
Average cloudiness, in tenths	4.1	4.2	3.1	4.6	..............Annual average cloudiness, in tenths, 4.0.
Rainfall	5.39	.18	3.98	14.08	Annual average rainfall, 23.63.

SACRAMENTO.

Mean temperature, maximum and minimum temperature, clear, fair, and cloudy days, average cloudiness, in tenths, and rainfall, by seasons:

SACRAMENTO.	Spring.	Summer.	Autumn.	Winter.	Annual Averages, Highest, Lowest, and Annual Totals.
Mean temperature	59.5	71.7	61.5	48.3	.... Annual average temperature, 60.2.
Maximum temperature	98.0	105.0	101.0	73.5	...Highest temperature, 105.0.
Minimum temperature	29.0	48.0	27.0	21.0	....Lowest temperature, 21.0.
Clear days	50.7	85.1	68.2	38.0	.... Annual number of clear days, 242.0.
Fair days	25.5	6.2	15.8	26.4	.... Annual number of fair days, 73.9.
Cloudy days	15.8	0.7	7.0	25.8	.... Annual number of cloudy days, 49.3.
Average cloudiness, in tenths	3.2	0.7	1.8	4.2	..............Annual average cloudiness, in tenths, 2.5.
Rainfall	5.22	.17	2.92	11.52	Annual average rainfall, 19.83.

RED BLUFF.

Mean temperature, maximum and minimum temperature, clear, fair, and cloudy days, average cloudiness, in tenths, and rainfall, by seasons:

RED BLUFF.	Spring.	Summer.	Autumn.	Winter.	Annual Average, Highest, Lowest, and Annual Totals.
Mean temperature	60.0	80.2	63.5	47.0	---------------------- Annual average temperature, 62.3.
Maximum temperature	101.4	110.5	106.0	80.0	.. Highest temperature, 110.5.
Minimum temperature	28.0	47.0	26.0	19.0	...... Lowest temperature, 19.
Clear days	·43.2	80.6	65.3	37.0	---------------------- Annual number of clear days, 226.1.
Fair days	29.7	10.3	16.6	27.2	---------------------- Annual number of fair days, 83.8.
Cloudy days	19.1	1.1	9.1	26.0	---------------------- Annual number of cloudy days, 55.3.
Average cloudiness, in tenths	3.8	1.0	2.1	4.5	---------------------- Average annual cloudiness, in tenths, 2.8.
Rainfall	7.08	.21	4.44	17.12	Annual average rainfall, 28.85.

OROVILLE, CALIFORNIA.

The following table of mean temperature, maximum and minimum temperatures, clear, fair, foggy, cloudy, and rainy days, and the rainfall for each month of the year 1885, at Oroville, California, was furnished by Mr. Hiram Arents, Voluntary Observer of the Signal Service, U. S. Army, viz.:

1885.	Mean Temp. at 6:30 A. M.	Mean Temp. at 2 P. M.	Mean Temp. at 8:30 P. M.	Mean Monthly Average Temp.	Maximum Temp.	Minimum Temp.	Clear Days.	Fair Days.	Foggy Days.	Cloudy Days.	Rainy Days.	Rainfall.
January	43.1	60.3	54.2	52.2	74	34	-----	-----	-----	-----	6	2.10
February	48.2	67.1	62.1	59.2	70	38	-----	-----	-----	-----	3	.73
March	53.1	72.2	70.1	68.2	86	46	-----	-----	-----	-----	1	.25
April	57.5	72.0	65.1	64.7	84	42	13	6	1	11	9	1.64
May	61.3	79.3	73.0	72.1	93	52	24	3	0	4	3	.65
June	62.2	80.1	74.2	73.0	90	57	21	4	0	5	2	.69
July	67.2	88.1	81.1	78.8	96	60	31	0	0	0	1	sprk'l
August	70.0	92.0	83.1	82.1	105	63	25	4	0	2	0	none
September	65.0	86.0	76.7	76.1	97	56	25	2	0	3	1	.20
October	59.1	81.0	69.2	69.2	94	52	21	4	0	6	1	sprk'l
November	53.0	62.2	56.1	57.1	74	42	7	2	2	21	18	11.27
December	47.2	61.2	52.0	53.0	75	37	15	1	3	15	9	5.53
Annual averages	57.24	75.12	68.07	67.16	86.75	48.25	*182	*26	*6	*67	54	23.06

* Total number of days for nine months only.

METEOROLOGICAL REPORT FOR JANUARY, 1884, 1885, 1886, AT OROVILLE,· CALIFORNIA.

By H. Arents, Voluntary Observer, Signal Service, U. S. A.

January, 1884.—Mean temperature for this month was 50.45°; the lowest temperature was 35° on the 18th; highest on the 5th, 70°. Maximum temperature for the month was 54.04°; minimum, 42.17°; but one frost this month, and very light, on the 18th.

January, 1885.—Mean temperature, 52.23°; the highest recorded for the month occurred on the 25th, 74°; lowest on the 24th, 34°. The minimum temperature as recorded for the month at 6:30 A. M., 43.09°; maximum recorded at 2 P. M., 60.26°, and at 8:30 P. M., 54.19°; light frost on the 24th.

January, 1886.—Mean temperature, 48.18°; the highest was recorded on the 27th, 66°; lowest on the 6th, 29°. Minimum for the month, recorded at 6:30 A. M., was 42.18°; maximum, at 2 P. M., 54.14°; and at 8:30 P. M., 48.22°. From the above comparison, last month was the coldest of the three mentioned. January, 1884, was 2.15°, and January, 1885, 4.07° higher temperature than 1886.

On the 1st of January a cold wave passed over California and continued fourteen days. During this cold spell the thermometer for each morning at 6:30 A. M. was standing on the 1st, 4th, 5th, 8th, 10th, 12th, at 32°; on 2d, 7th, 31°; 3d, 34°; 6th, 29°; 11th, 40°. Ten of these days ice formed from a sixteenth to quarter of an inch in thickness. In the orchard and nursery of Gardella Bros., of Oroville, out of one thousand yearlings and five hundred three-year-old Lott's seedling orange trees, not one of them was injured by this severe test, notwithstanding every morning they were heavily coated with white frost. Mean barometer for this month was 30-02.50°; the extreme was, on the 2d, 30.36°; and 29.42° on the 18th. Prevailing wind was southeast; 15 days southeast; 12 northwest; 2 west and 2 east; 15 days cloudy; 11 clear; 2 fair; 3 foggy. It rained on the 13th, 14th, 17th, 18th, 19th, 20th, 22d, 23d, 25th, 26th, and sprinkled on the 12th; rainfall for the month 5.17 inches. From July 1st to December 31st, rainfall was 17 inches; total to February, 22.17 inches. From July 1st to December 31, 1884, the ·rainfall was 13.73 inches, and January, 1885, 2.10 inches. Total last season to date, 15.83 inches. Excess of the present season over last, 6.34 inches.

STORMS ON THE PACIFIC COAST OF AMERICA.

[From the Annual Report of the Chief Signal Officer.]

The storms of the Pacific Coast most resemble those of western Europe, than the storms which frequent the eastern coast of the United States. The latter move littorally, and follow a northerly and easterly course under the thermo-dynamic influence of the Gulf Stream and the mechanical agency of the great southwest equatorial current of atmosphere, which remarkably coincides with the oceanic Gulf Stream.

But on the Pacific side of our continent, the storm-controlling forces act in a direction from west to east, especially upon the coasts of

CALIFORNIA,

Oregon, and Washington. The warm Kuro Siwo, or Japan stream of the great ocean, after reaching the middle latitudes, on the way to the Aleutian Islands, is superficially brought under the propelling power of the westerly or anti-trade winds, and a large drift of this Pacific Gulf Stream is borne eastward as a decidedly marked warm stratum of surface water, and strikes upon the western shores of America nearly at right angles. This agency, as well as that of the general atmospheric movement on our Pacific Coast, serves to give character and direction to the storms and cyclones which reach it, no doubt, from the western Pacific Ocean.

From San Diego to the Straits of Juan De Fuca, from December to April, the storms of the

PACIFIC COAST

Set in, with southeasterly winds, veering as the storm center progresses, to southwesterly. The closing winds from the north of west are very severe, and, as they blow onto the lee shore, are to be apprehended by vessels, even though in port. Instances are not wanting in which vessels have been sunk in the Pacific ports of America by these gales from the west. These southeasterly gales are more frequent and violent north of San Diego, and thence along the coast to

BRITISH COLUMBIA.

This can be easily understood from the fact, as established by Blodget, that the humidity and rainfall of the region stretching from San Francisco northward to Vancouver Island are nearly three times as great as of that south of San Francisco. Unless forced by other causes to deviate from the regions of greatest humidity, we know storms seek or are drawn into such regions for their necessary supply of aqueous vapor. On the Pacific Coast there are no other known agencies which would cause such deviation. It follows, therefore, that the ports of

SAN FRANCISCO,

And Portland, Oregon, and the waters of the adjacent sounds, are more endangered by storms than San Diego or those points along the coast between San Diego and San Francisco. In Summer the latter port is so far south of the usual storm track that it is comparatively safe; but it is otherwise from December to April. The northeasterly wind, which on the Atlantic seaboard is often a violent premonitor of a storm, on the California coast and northward does not precede, but follows the cyclone in its closing northwest quadrant, and is usually of moderate force.

After striking the Pacific Coast the storm will generally advance with but little diminution of cyclonic intensity, but with diminished progressive motion, in a direction east-northeast. The violence of the storm will not cease till the center has passed beyond the

COAST RANGE MOUNTAINS.

The great upper current or stratum of warm and moist equatorial atmosphere, which in England has been observed to move in a southwest direction, is on our Pacific Coast less meridianal in its course, and pushes more toward the east, especially north of the parallel of 48° north latitude, where it is favored in this more easterly direction by the orographic features of the continent, which are less elevated and bold than they are south of this parallel. Vessels sailing south from San Diego to Mexican ports are peculiarly exposed from June to November from severe gales, beginning generally at southeast or southwest. These southeast gales may be looked for in still greater severity and frequency, especially during Winter and the equinoctial seasons, all the way from San Diego to the Straits of

JUAN DE FUCA,

And attended with thick, rainy weather. Vessels sailing northward to San Diego from Mexican and southerly ports, should they encounter a gale moving up the coast, should stand off for the starboard tack, thus getting the eastward winds of the cyclone, which removes the danger of going ashore.

RECURVATION OF STORM–PATHS IN THE EASTERN PACIFIC.

Mr. William C. Redfield, on cyclones on the Western Pacific, says: Most of the cyclones which I have last described, however, must have been recurvated in a more advanced position in the Pacific Ocean; and in their subsequent northeasterly progress they would fall almost perpendicularly upon the coasts of the two Californias, or the more northern Territories. Thus, instead of sweeping a great length of these coasts successively, as happens on our Atlantic border, these cyclones appear more like local storms, and can not be traced consecutively on the coast line. At the point of intersection with the coast, the first and main portion of the gale will be felt from the southeast, on the center path, or more southerly in the right hand quadrants; and near the coast the northeasterly or reflex winds of the cyclone, pertaining to its first left hand quadrant, will not be strongly developed.

From Cape San Lucas, 23°, to San Diego, 32° north latitude, the coast is subject to violent gales from the southeast from November to April, and that they are more frequent as we go toward San Diego. Before their recurvation these cyclones are likely to have passed westward in lower latitudes than those which fall on the Mexican Coast.

From San Diego to San Francisco the coast is subject to southeasterly gales, like those of the coast of Lower California, but they are more frequent here, and blow with greater force. These gales last from twelve hours to two days, and are accompanied by heavy rain, which lasts till the wind changes, which it often does very suddenly, and blows as hard for a few hours from the northwest, when the clouds clear off and fine weather again succeeds. This is a clear description of the phenomena of cyclones, as shown on their center paths, while moving in a northeasterly course.

From San Francisco to the Straits of Juan De Fuca, hard gales from all points of the compass may be looked for at all seasons. These begin generally from southeast to southwest, bringing thick rainy weather with them. After blowing from these quarters for some hours, they fly round to the northward, by the west, with little if any warning, and blow even harder than before. These changes show the observer to have been in the right hand quadrants of the gale, as most often will happen, and are but counterparts of the changes met with in the cyclones encountered in the same latitude in the north Atlantic.

THE NORTHERLY WINDS OF CALIFORNIA.

By J. H. C. Bonté, Secretary of the University of California.

A frank and fair discussion of the northerly winds of California is much needed. The first necessary step in this investigation is a correct knowledge of the topography of the valley in which these winds prevail.

TOPOGRAPHY.

The great central valley of California, known under the names of Sacramento and San Joaquin, heads at Redding in the north, and extends to Tehachapi Pass, in the south—a distance of four hundred and fifty miles, with an average width of about forty-five miles.

The Sierra Nevada Range of mountains bound this valley on the east with a continuous wall, which has an average elevation of about six thousand feet, and an average width of about eighty miles. This range is well timbered from the foothills to the summit. It has a heavy Winter snowfall, which remains on the highest points during the whole year, and feeds the valley rivers during the Summer. There is no Summer rainfall on this or any other mountain range in California.

The valley is bounded on the west by the Coast Range mountains, a range with but one gap—that at San Francisco—which is about half way between the north and the south end of the valley. The average elevation of this range is about two thousand five hundred feet, and its average breadth about forty miles. This range is poorly timbered on its eastern slope, is rarely covered with snow, and then only for a very short time. The winds of the Pacific Ocean pile up great fog banks on the western slope now and then, keeping the air on the ocean side moderately cool.

These two ranges are united at the north end of the valley by other ranges, which are thus described by Rev. E. L. Greene: "There are several mountain ranges in the north end of the State running in different directions. The Siskiyou Range, which is largely in the State of Oregon, runs mostly east and west, and averages about eight thousand feet in altitude. From Mount Shasta, running in a northwesterly direction, is a high range, separating Shasta Valley from the lava beds. Another high range runs from the southwest base of Mount Shasta in a westerly direction. Between these more conspicuous ranges are lower ranges, cutting up the whole region in a succession of mountains and valleys. The mountains, on their northwest slopes, are here and there densely timbered, at an elevation of three thousand and four thousand feet. Below three thousand feet, the southward and eastward slopes are naked, or clothed only with chap-

face is principally covered with decomposed granite. This great valley is separated from the Bays of San Francisco and San Pablo by the Diablo Range, which extends from the Straits of Carquinez in a southeasterly direction about one hundred and fifty miles, where it terminates in low ridges running out into the San Joaquin plain. This range has an average elevation of about two thousand feet, an average width of about twenty miles, and is poorly timbered; indeed, almost nude.

Livermore Pass, in this range, has an elevation of six hundred and eighty-six feet, and Pacheco Pass, one thousand four hundred and seventy feet.

The trend of the valley, and all its walls, is southeasterly.

THIS LONG VALLEY, THUS WALLED IN,

Is veined by several considerable streams: the Sacramento, Pitt, Feather, Yuba, American, Cosumnes, Mokelumne, Calaveras, San Joaquin, Kings, White, and the Kern, all of which come out of the Sierra Nevada Mountains. The two great rivers, the Sacramento and the San Joaquin, head in opposite directions, but approach and unite at Suisun Bay, through which they empty into the Straits of Carquinez at San Pablo Bay. The Coast Range mountains contribute but little water to this valley, save in the Winter. The south end of the valley contains two or three small lakes, and several of the rivers are fringed by inconsiderable marshes. The valley is nearly level throughout its length, and has an elevation of about two hundred feet. This even surface is broken only by the Marysville Buttes, which rise abruptly out of the plain. This is a range of volcanic hills about six miles wide and twelve in length, with three peaks. The valley is generally bare of native trees, though the oak grows to some extent throughout. The substratum of the valley is a deposit of gravel and sand, with a depth of about two thousand feet. The surface soil is sand and dissolved volcanic material, mixed with vegetable mold. This great underlying bed of sand and gravel is always thoroughly saturated with water to within a few feet of the surface. The Winter rains saturate the surface soil until the two moistures meet; but the Spring and Summer evaporation dries the soil to an average depth of about two feet, leaving the upper surface cracked, dry, and hard, drying every kind of vegetation that does not extend its roots to the moisture below, or receive water from irrigation.

The soil at the south end of the valley is very largely composed of sand, gravel, and hardpan, substances capable of absorbing and containing vast accumulations of heat.

We have now before us a general view of the topography of the valley in which the northerly winds move.

In studying the meteorology of this valley, we would naturally expect the rarification of the valley atmosphere to draw in the cold air from the Pacific Ocean, through the Golden Gate, the Diablo passes, and the Straits of Carquinez.

But the Diablo Range, the narrowness of the Golden Gate and the Straits of Carquinez, and the elevation of the Diablo passes, seem to prevent the extensive and ready influx of ocean air, until certain contingencies occur, which enable the ocean winds to overcome these obstacles. In the meantime, the only winds that visit the valley are the northwestern winds, of which we are now writing. These northern winds are felt throughout the valley, and to some extent in San Francisco, and in Napa Valley. As a general rule, they are more virulent in the extreme north and the extreme south end of the Sacramento (and San Joaquin) Valleys.

The atmosphere from the Golden Gate, and the three bays, seems to modify the wind, ordinarily, in the center of the valley. The general course of these currents of heated air is, of course, determined by the trend of the valley and its walls.

Without attempting an accurate statement of the periodic recurrence of these winds, we may say that the ordinary current returns about once in two weeks. Hot blasts occur about ten times during the year, while the extremely vicious blasts occur only once in six or eight years. We are, however, without data for the settlement of these points, unless we can find the needed information in the meteorological records of the Central Pacific Railroad Company. Their continuance is from one to thirteen days, and the average about three days. Ordinarily, the wind ceases at night, and is followed by a slight breeze from the south, though frequently the movement continues all night with considerable force. Counter currents are not generally noticeable during the prevalence of the north wind.

THE ORIGIN AND CAUSE OF THE NORTHERLY WINDS.

In discussing this point, we must distinguish between the cause of these winds and the cause of their disagreeable characteristics. If we are to regard the northern winds as special or local winds, we must first seek for local causes. Wind is, of course, a certain quantity of air set in motion by a change of equilibrium, and this loss of equilibrium is produced by the rarification of the atmosphere. The sun's rays penetrate the atmosphere at the south end of the valley, and being incapable of returning to celestial space through the same medium, they leave their heat in the sand and decomposed granite, where it is entrapped and stored. This accumulated heat rarifies the air, which ascends and creates a practical vacuum at the south end of the valley. The contiguous air to the northward then rushes in to restore the equilibrium, while the heat in the soil creeps northward until the whole surface of the valley becomes heated, when we have a practical vacuum four hundred and fifty miles long, with an average width of forty-five miles. The air north of the Sacramento end of the valley then rushes into this long vacuum, which is more perfect at the south end, and the result is our northerly winds. Considering the northers as local winds, this is the only explanation that science has to offer as to their origin and cause. But, as the science of meteorology advances, we are led to look for more general causes—causes connected with the general circulation of wind currents.

For a long time the cold southwest winds of France were attributed to local causes, but more thorough investigation of barometric pressures in Spain, France, and Italy, during the prevalence of that wind, established the fact that it belonged to a broader system of circulation. And it is probable that more extended observations of barometric pressures will connect our northers with a wider system of atmospheric circulation. As having a possible bearing upon this point, we note the fact that the north-westerly winds of this valley are sometimes almost simultaneous with the northwesterly winds of the regions about Santa Barbara. Hittell, in his "Resources of California," gives an account of two such nearly coincident currents occurring at Stockton and Santa Barbara in June, 1859. It is also to be noted that, in the Winter season, the southeast wind, which brings on rain, is preceded by the northerly wind, carrying southward the rain clouds at a very high altitude. Further discussion on this point must

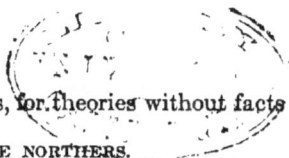

be postponed until we can accumulate facts, for theories without facts only confuse.

CHARACTERISTICS OF THE NORTHERS.

First—The northers are cool in the Winter and early Spring; hot and dry during the Summer and Fall; a very wet Winter, however, postpones the high temperature until about the middle of June.

Second—The speed of movement is sometimes very great, approximating the rapidity of a gale.

Third—The evaporating power is very considerable at all seasons, though greatest in the Summer and Fall.

According to my own observation the north wind will sometimes evaporate from a glass goblet a full inch of water in twenty-four hours, while a south wind of equal force will not evaporate to an appreciable extent.

Fourth—The atmosphere during the prevalence of a Summer and Fall north wind is usually free from clouds, though now and then a very few thin streaks of cirrus clouds are visible. The presence of distinct clouds in any quarter is always prophetic of a subsidence of the northers.

THE BAD EFFECTS OF THE NORTH WIND.

I desire to put upon record a brief statement of some of the evil effects of the north wind, hoping that subsequent and more careful investigation may enable me to modify these statements.

First—The effect of prevailing northers upon the vegetable economy. These evil effects are more noticeable in Spring time when everything is tender and full of sap, and less observable in the Fall when the fibers have become tough. A heavy Winter rainfall which saturates the earth sufficiently to postpone complete evaporation, protects vegetation until it is strong enough to defend itself, for water seems to neutralize the wind's evil power. After a dry Winter the north wind becomes extremely prejudicial to some forms of vegetation. These evil results are, however, modified by protecting houses, hills, fences, and trees.

Dr. Harkness reports that the officers of the United States Army, at the Presidio, succeeded in protecting their gardens by very simple fences; a defense which would prove insufficient further north, away from water. Yet it happens that the side of a tree next to the wind is injured, while the opposite side remains unharmed. But I prefer to give the language of practical observers, remarking that the facts now to be mentioned occurred only during the severest blasts.

Mr. Hoagland—Apples are sometimes baked or burnt on the trees before they ripen, on the side toward the wind. This spot becomes hard, and a dry rot sets in. The rest of the apple ripens, but this spot remains hard and woody.

Josiah Johnson—In April, 1859, twenty miles south of Sacramento, near the confluence of the Cosumnes and Sycamore Rivers, the north wind continued nine days. Young rose and sycamore leaves were blackened and charred, curling up like burnt paper. A few days ago, and during a north wind, I plucked from my yard some flowers usually very fragrant, and found them to be void of their usual perfume. I took them into the house away from the wind, and their fragrance returned at once. This experiment has been frequently repeated, with the same uniform result. The leaves of the acacia tree, which close up only at night, close up soon after the beginning of a north wind.

Thomas Sayles, for twenty-five years a practical nurseryman, twelve years of which was spent in California—I have known the young sprouts on cherry and peach trees to be killed perfectly dead in two hours. I have seen this frequently. Evergreen trees, when taken up, and while the roots are well bagged and watered, are often killed before reaching a near market, even in an ordinary north wind. The effect is the same as fire. I have known apple trees ten years old to be killed by a few days of north wind. It cuts down young orange trees like fire or frost; and we dread the north wind more than we do frost.

Colonel Wilson, of Nord—The north wind nearly destroyed the grain crop in 1875.

William Gwynne—I saw in 1851, Mr. Harbeson's wheat field in Yolo County, which was then in the milk, wholly blighted in three days.

Miss Brewster—I have seen the sulphur colored rose, when in vigorous bloom, turn black in three quarters of an hour; a blanket covering furnished no protection whatever. Gelatinous water flowers are not noticeably affected. Hyacinths lose their fragrance in the north wind.

General Cadwalader—Potatoes are not seriously injured while the ground is moist; but later in the season, if not irrigated, they are seriously damaged. Walnut trees are sometimes burnt on the north side from branch to root.

EFFECT ON THE ANIMAL ECONOMY.

Cows give fully one third less milk during the prevalence of a north wind. Horses have no travel in them during a north wind, and seem to lack breath, and require double the usual effort to do their ordinary work. The north wind frequently kills young turkeys and chickens. In a very few hours they sicken, begin to droop, and die. Careful people cover them during a north wind. Cattle always become thirsty, nervous, and restless.

Mr. Bassett—Birds generally cease to chirp and sing; seem to be feverish, and lose their appetite. They bunch up as in cold weather, and refuse to bathe. These effects are as noticeable when the north wind is cold as when it is hot. Setting hens become nervous on the nest, and get off more frequently than at any other time. The nose of the setter dog grows dry and warm, and they will not take the scent.

William Gwynne—In 1851 I was traveling in Yolo County during a north wind, and saw little birds fall dead from the trees. I took one almost dead to a spring under a shade of grapevines, bathed and fanned it until it came to life. It would not move from the shade.

General Cadwalader—Coveys of birds are sometimes killed. All animals seek the wells and springs during the north wind. Sheep sheared during the north wind lose greatly in weight.

EFFECT ON MAN.

Dr. Harkness—Healthy and strong individuals feel an inconvenience, an agitation, a heaviness difficult to express; the muscular system is more sluggish; individuals afflicted with rheumatism feel their pains renewed; neuralgias increase in intensity or their paroxysms reappear; men are cross-grained and quarrelsome; fights are of frequent occurrence, and our landladies are seldom found in their usual amiable mood. In general, our patients afflicted with chronic or acute affections feel an aggravation of their principal symptoms; they are more fatigued, more agitated, and their febrile state is increased, while, without being able to give any reason for it, they are often gloomy and despondent. Urinary secretion contains an excess of solid ingredients, and is diminished in quantity. The functions of the brain are also disturbed by the same morbific influence, resulting in slight headache and drowsiness, with marked disinclination for either mental or physical action.

I add the testimony of Dr. J. S. Cameron, of Red Bluff: The north wind produces a feeling of depression and nervous irritability; the lean and spare made people being less susceptible than persons of a corpulent habit. The first effect consists in a feeling of tightness in the respiratory organs, often associated with headache; a dryness of the skin; thirst, and a diminution of the excretions. The majority are made pale by the hot winds of Summer time; the eye will generally show signs of congestion, and the after effect usually increases the determination of blood to the surface. Irritability is coincident with the north wind, caused, no doubt, by the general atmospheric disturbance. The north wind causes an increase in the amount of liquids drank. The hair becomes dry and crispy because of evaporation. Consumptives are made very much worse; they are very much prostrated by it, and in fact must leave the valley during the prevalence of the north winds. All diseases of the respiratory organs, except asthma, are made worse. Neuralgia is also usually aggravated. The prevalence of a north wind of long duration in the Winter and early Spring is uniformly accompanied with or succeeded by an epidemic of pneumonia; the one in January, 1873, having produced over sixty cases in Red Bluff; the mortality, however, was small. Rheumatism may, in some cases, be benefited, but I have no recollection of any case where it was; but unquestionably it is less prevalent during the continuance of the north wind. Those who suffer from asthma are singularly free from it, but I attribute this to the fact that the climate of Red Bluff is a specific for the cure of that disease.

Dr. Allendorf, of Red Bluff—The complexion during a north wind is apt to become sallow, rough, and dry; some become very pale, others ruddy. The hair becomes dry and rough, and the wind has a burning and blinding effect on the eyes; also produces headaches and sleeplessness. The young do not seem to suffer as much in proportion as those of forty years and over, but all dread and dislike it. The north wind has a very perceptible effect on persons, especially after middle life. In those exposed there is a sense of lassitude, pain of the joints and limbs, disinclination to exertion, restlessness. The secretions of the mucous membranes are much decreased. There is in man a shrinkage of weight as great as one pound per day. A considerable number of intelligent persons have, by frequent experiments, come to the conclusion that the human body loses by evaporation, during a strong north wind, from one to two pounds per day. But this point needs more careful investigation.

The late lamented Dr. Ed. M. Curtis (Sacramento), a man distinguished for his correct and close habits of observation, gave me, shortly before his death, the results of his observations on the effects of the north winds on his own person. He suffered for years, and finally died of consumption. He said that the north winds were to him exceedingly enjoyable, and that he felt better during a north wind than at any other time. While engaged in studying this subject, I have found well people who claim that they feel a happy exhilaration during the prevalence of the north wind. Among these are persons of every variety of temperament. Nevertheless, it is an established fact that many disagreeable results come from our north winds; and the question arises whether we can account for these disagreeable characteristics and bad effects of this so called "poison wind."

HYPOTHESES. ·

Hypotheses are imaginative efforts to overcome difficulties, and their use is fully justified by experience.

First—It is claimed that the heat and dryness of the north wind are communicated to it by the dry plains and stubble fields of the Sacramento Valley; and in support of this hypothesis it is said that the north wind does not become peculiarly vicious until after harvest. The objection to this hypothesis lies in the supposed fact, probably true, that the north wind is more vicious just where it first touches the Sacramento Valley than afterwards, and that its exasperating qualities decrease as the current moves southward. As my personal observations of this wind have been confined to Sacramento and the Bay of San Francisco, I am not able to determine the question at issue. One thing, however, is certain: I have received descriptions of the evil effects of the north wind as far south as Stockton, which could not be surpassed by any similar occurrences at the north end of the valley. But no matter how much truth this hypothesis may contain, it does not explain the cause of the exasperating characteristics of the north wind as felt at the extreme north end of the valley.

Second—Dr. Cameron states another hypothesis: "The heat of the north wind in Summer time seems to be communicated to it from the lava beds of Northern California, as I am informed that above Yreka they begin to be pleasant winds, even in the hottest Summer months." The Rev. Ed. L. Green, of Yreka, says: "We have north winds, though no high winds from that quarter. They are cold, bringing frosty nights, sometimes even in June. Later, after the warm weather sets in, they effect an agreeable change in the temperature. We have no wind here corresponding to the dry, disagreeable north wind that blows down the Sacramento Valley." As Yreka lies due west of the lava beds, so as not to be influenced from that direction, the second hypothesis may yet be established.

Third—A third hypothesis is offered. In this it is claimed that the wind which sets into the Gulf of California passes up through the arid plains of Arizona northward, curving westerly and then south, entering the head of our great valley; and that the heat and other disagreeable qualities of the north wind are derived from the plains of Arizona and intervening deserts. Of this hypothesis it is sufficient to say that we are not yet in the possession of established facts with which to sustain it in a scientific manner.

Fourth—I think we will make better progress by separating the subject of causation, and by simply considering, first of all, the cause of the dryness of the north wind, without any reference to its other disagreeable characteristics. The dryness of this wind is partially explained by the fact that it passes over lava beds and dry mountains. Indeed, during the Summer all the mountains of California, even where they are not denuded, are gigantic dust heaps.

Fifth—It is probable, however, that the northerly winds have their origin in the far western Pacific Ocean, and it is claimed by observing travelers, that they are exceedingly dry in Summer time, far out from our coast.

Sixth—But all dry winds are not necessarily evil in their effects. We have, then, to account for their disagreeable and pernicious qualities. And for this purpose we resort to a sixth hypothesis, which depends for its support chiefly on the dryness of the north wind. For the sake of convenience, we call this the electrical hypothesis. Electricity is capable of being massed, condensed, rarified, and also of discharging itself. It exists in positive and negative forms, in every object upon the earth; and negative electricity is just as active and efficacious for all practical purposes as the positive. The earth is practically an infinite reservoir of both electricities; though by comparison the earth may be supposed to contain, on the whole, negative, while the atmosphere is charged with positive. When the normal relations between the earth and the air are undisturbed, there exists an easy, natural, and imperceptible interchange of electricities, which preserves the general equilibrium. The north wind, being the most perfect insulator and best non-conductor, necessarily insulates the earth. In this condition, the earth no longer receives electricity from the air, for this dry wind can neither give nor receive. The result is, that the surface of the earth, and everything upon it, is excessively charged with an imprisoned electricity. It is believed that this hypothesis is founded upon well established and clearly defined principles. But common sense requires something more, and demands other proof of the presence of electricity at these times.

Dr. H. W. Harkness remarks in his essay upon this subject: "We feel in a north wind, sometimes, as when we receive a moderate shock from an electro-magnetic battery."

Dr. W. R. Cluness—I have frequently noticed, after riding in the north wind, that my hair became dry and stood out. Running a comb through produced the electric snaps.

Matthew Cooke—I have, after driving in a prevailing north wind, put my finger to the belt of my driving-wheel, which drew from my finger nails a steady blaze two inches in length.

Mr. Hoyt—During the north wind the tails of my horses sometimes stand out fan-like. The use of the comb and brush produced the electric snap.

It is evident, therefore, both from these simple facts and the principles of electricity, that during a north wind everything connected with the earth is insulated and heavily charged with electricity. It has, however, been claimed in support of the theory, that the north wind imparts electricity from itself, and from the upper regions; that a stove insulated from the earth by vitrified bricks was so heavily charged with electricity as to impart a very heavy shock to one who attempted to handle the stove.

But in opposition to this, it must be admitted that the original instance referred to was never examined; that all similar experiments have failed to produce this result, and that the proposition is highly improbable. We may, however, imagine a stove placed in a very damp place, where there is damp air sufficient to conduct electricity to the stove from the ground; but we can hardly imagine that the stove receives electricity from the north wind, though the friction might generate it. If I have established this member of my hypothesis, namely, that during a north wind every object is insulated and heavily charged with electricity from the earth, we are prepared with a reliable explanation of many of the results of the north wind.

We put the statement in several forms: Tyndall says: "When an electric current encounters resistance, heat is developed. This heat is sometimes so intense as to reduce metals to a state of vapor." This being true, the excess of electricity in plants and animals which always seeks to reëstablish its equilibrium meets with resistance in the north wind at the surface of the object and hence an extraordinary degree of heat. Again: It is supposed that a non-excited body contains an equal amount of negative and positive electricity. Ordinarily this is the condition of objects on the surface of the earth, and the result is a neutral state. But friction decomposes these two into one or the other of the two elements, and the result is action.

Now, then, the north wind finds the animal and vegetable economy charged with negative and positive electricities in a neutral state, and the friction of the wind decomposes the two elements, producing a marked disturbance of electricity, and this disturbance is quite sufficient to account for all these effects of the north wind that are not accounted for by the simple dryness of the air.

But we must add the further fact that there exists an electric current in all animals and vegetables; that there is a current perpetually circulating between the internal and external portion of the muscles of the animal. This animal electricity, no doubt, derives its source from chemical action, constantly in progress, in connection with the vital processes. But this chemical process must, more or less, be interfered with by the disturbances produced by the north wind.

This fact alone will account for very many of the evil results of the north wind, both in the vegetable and animal economy. To say the least, this hypothesis accounts for the intense heat and the nervousness felt by those who are susceptible to this malady. It will also account for the exhilaration felt by others.

Seventh—We venture still another hypothesis: The science of chemistry has demonstrated the existence, in the air, of chemical elements, such as oxygen, nitrogen, aqueous vapor, carbonic acid, ammonia, iodine, and ozone, elements that are perfectly harmless in their normal combinations.

But the north wind may enforce different combinations, productive of great temporary discomfort to man. Future and more thorough investigation may find in this hypothesis a suggestion of considerable value. The discoloration of the sulphur-colored rose is suggestive of some chemical action in the north wind. Several other hypotheses might be offered, but those already named are sufficient for our present purpose.

ECONOMIC VALUE OF THE NORTH WIND.

Heretofore, in this discussion, we have assumed only the harmfulness of the north winds. But are they wholly valueless? Have we a right to assume that these natural currents are only evil, and evil continually? The constant circulation going on in the atmosphere renders impossible the entire consumption of any substances necessary to maintain the life of organized matter—such as oxygen, aqueous vapors, etc.—and it also prevents any dangerous accumulation of deleterious matter—such as carbonic acid. The existence of animated nature is intimately connected with this circulation. Thorough investigation will establish the fact beyond a question that the north winds are of inestimable value to the great central valley of California. And I am confident that the ordinary estimate of the north wind by the people of this valley is an exaggeration of its disagreeable qualities. The majority of well people cannot tell by their own feelings, without external observation, whether the wind, at a particular time, is from the north or the south. I have often heard men of well disciplined minds, who considered themselves particular victims of the north wind, complain bitterly of a north wind when the wind was directly from the south. Fully one half of the misery attributed to the north wind is purely imaginary, or the result of indigestion or indolence, or the simple result that follows all atmospheric disturbances. The people of this great valley have, in an unconscious and imitative manner, agreed to consider themselves miserable during a north wind. The psychology of this morbid condition would be of interest, but lies beyond the line of our present discussion.

For the purpose of opening the subject for further consideration, I now offer a series of suggestive propositions, and if, in the statement of these, I make some use of the imagination, it must be remembered that science regards a legitimate use of that faculty as of preëminent value. Knowing that any exuberance of the imagination in this connection will meet with remorseless punishment, I shall restrain this faculty within just limits.

First—The peculiar, dry, and moderately exhilarating climate of this great central valley is a result of the northerly winds. Without this evaporating power, the valley; its atmosphere and its very walls, would drip with perpetual moisture; pernicious fogs would cloud the sun and conceal the valley, with no possibility of escape from these walls; and the result would be a humid, relaxing climate, susceptible of that high degree of heat not marked by the thermometer, but felt by the system. Then eighty degrees of heat would be the equivalent of one hundred under present circumstances.

Second—Without the north winds, and with the increase of moist heat, the vegetation now cultivated, and so highly prized, would be overlapped, overwhelmed with gross tropical growths. The exceeding fertility of the soil would crowd and cram the soil with excessive growth. It is not difficult to see the force of this proposition, in view of the well known fact that judicious and careful irrigation and culture will, even now, produce a forest of fruit trees, of vines and plants, within a period of five years. As matters now stand, we can select and cultivate any or all of the products of the various zones. Between our present happy condition and the wretchedness of a purely tropical state, lies our only defender—the north wind.

Third—As a natural and necessary sequence to our first two propositions, there comes the third—the north wind, by its desiccating power, is a preventive of disease. By the north wind, excessive growth, and therefore excessive decay, and therefore excessive malaria, and therefore disease—all are prevented. Without the north wind, ague and the virulent fevers would prevail universally and at all times. It is also within the range of possibility that we are indebted to this agency for our comparative exemption from sunstroke and hydrophobia; at least, it is proved that sunstroke occurs only after very wet Winters. It is but reasonable to believe that this desiccating power, which prevents and dissipates the noxious exhalations of animal matter, defends us against all those diseases that are propagated by poisonous pus.

Fourth—The north wind possesses curative powers. This proposition is rendered probable by the curative effects of similar winds in other countries. The harmattan wind of Africa, which possesses the same characteristic as our north wind, is preëminently curative. Intermittent fever is cured by the first breath of that wind, and remittent and epidemic fevers disappear as by enchantment, while infection of all kinds, including the artificial infection of vaccine virus, fail during the prevalence of that wind. The natural presumptions of the case favor the truthfulness of this proposition. I am satisfied that surgical treatment in this valley is rendered more easy than in moist climates. If so, it is because of the curative qualities of our dry climate, which is the natural product of our north winds.

The treatment of disease by electricity is a department of medical science which is yet in its infancy, with all the probabilities in its favor. It seems plain to me, admitting the value of medical electricity, that Nature, in this valley, is already administering this curative agency, in a manner already explained in our fifth hypothesis. It seems probable, in view of that explanation, that the medical faculty has it within reach to control the natural results of the presence of an excess of electric fluids. If the excess of electricity comes from the earth, and not from the air, the amount of electricity in each patient may be controlled by still further insulation and discharge. This further insulation from the earth may be accomplished by glass under the posts of the bedsteads of the bedridden, or by encasement of those who move about in silk underclothing. I am told by Dr. Harkness that the last expedient is frequently resorted to in similar winds of India. The medical fraternity are under obligations to humanity that ought to lead them to important results in this direction.

Fifth—The north winds, following the rainy season, by drying and baking the soil, dissolve and pulverize its particles, thus freeing its inherent productive powers. A similar result is produced in colder climates by the alternations of rain, frost, and heat. This line of investigation is commended to intelligent agriculturists.

Sixth—The short, dry, seedy grass upon which our farmers rely during the Summer and early Fall, and which is so quickly destroyed by moisture, is cured and preserved by our north winds. This short grass, at the proper moment, is seized by the north wind and quickly cured; and cured in a way that preserves all its nutritive qualities from evaporation.

Seventh—Fineness of fiber and concentrative nutriment is imparted to all our vegetable growth by the north wind. And it is possible that the grape and strawberry may receive their delicate flavor from the same source. At least we are sure of this: that without the north wind exceeding grossness would characterize all our vegetation.

Eighth—The north wind, while it sometimes destroys, often brings our cereals to a rich and profitable maturity; imparting to the berry a solidity and flintiness that enables it to resist the damaging effects of moisture. How far wheat is indebted to the north wind for its glutin and thinness of husk I am not able to say, but I believe, from the few facts in my possession, that it performs a kindly office in this direction. I am convinced that the absence of the north wind, and the inevitable increase of moisture, would give us mere bigness of berry, to the sacrifice of flavor. Without the north wind our grain would naturally continue to grow a month longer, during which time it could only increase in size. And, as there is a circulation of electrical currents in all fruit, vegetables, and grain, it may yet be found that the electric disturbances produced by the north winds are extremely favorable to all our crops. The north wind protects our crops from destruction by animal and vegetable parasites. Our comparative exemption from the ravages of weevil doubtless arises from the desiccating power of the north wind, and perhaps, in part, from the electric conditions. The most common enemy of the vegetable economy is the fungi. Fruit trees are injured by microscopical fungi: potatoes, onions, lettuce, vines, hops, peas, cabbage, and turnips, each have their inimical fungi. In moist climates the grain crop is smitten with red-rust, mildew, smut, and perhaps other fungi. The conditions favorable to the growth of fungi are

Ninth—The economical value of the north wind is discernible in its power to preserve the various materials useful and necessary in our civilization. The dry air and its intense drying influence must necessarily perform an important office in preserving from rapid decay fences, barns, houses, railroad ties; and I am confident that the same influence must protect iron in every form from destructive rusts. Facts sufficient to establish this proposition are doubtless within our reach.

Tenth—The desiccating power of the north wind, which enables us to give all wood a thorough seasoning, will lead of necessity to the establishment of extensive manufactories of wood throughout the valleys. At present this is as much a prophecy as a proposition.

Eleventh—The wind now under discussion facilitates practical telegraphy, by giving, what is rarely attained elsewhere, a perfect insulation. Upon this subject all practical operators are agreed.

Twelfth—It is within the power of ingenuity and industry to control and utilize these winds.

We have already discovered that vegetation can be measurably protected by trees, houses, fences, and water. When this valley becomes fully settled, and the great farms are broken up into small homesteads; when there shall be fifty thousand houses, orchards, and forests where there are now ten; and when a judicious system of irrigation shall be adopted throughout this great central valley, then the vicious qualities of the north wind will cease altogether. Indeed we only fear that the future development of the valley may deprive us of even the advantages derived from this wind. The results of the Suez Canal are suggestive in this direction. Formerly rain was unknown on that part of the Red Sea, but since the building of the Suez Canal showers have fallen regularly about once a fortnight. The result has been to start vegetation, even on the Asiatic side, in the most wonderful manner, and if things go on as they have begun the sands of the isthmus will be covered with forests in another half century.

From all I can learn the north winds have lost much of their violence during the past twenty-five years. If so, we may assume that the development of the country has already begun to change the character of these winds.

PALESTINE CLIMATE VERY SIMILAR TO THAT OF CALIFORNIA.

SIROCCOS WHICH PRODUCE MORE ILLS THAN ALL THE NORTH WINDS OF CALIFORNIA—THE RAINY SEASON AND THE DRY SEASON IN THE HOLY LAND.

An interesting report of the climate of Palestine, by Mr. Selah Merrill, United States Consul at Jerusalem. The following extract will be found very interesting reading for Californians:

SEASONS IN PALESTINE.

There are in Palestine two seasons, a rainy season and a dry one. The shortest rainy season in twenty-two years has been one hundred and twenty-six days, and the longest two hundred and twenty-one days, while the mean duration of each season has been one hundred and eighty-eight days. On the other hand, the shortest dry season for the same period was one hundred and thirty-four days, and the longest two hundred and eleven days, while the mean duration of each was one hundred and seventy-seven days.

COMMENCEMENT OF THE RAIN.

The time of the commencement of the rain is uncertain, and varies many weeks between the two extremes. In September the people begin to talk about rain, and to look for the token of it, but rain very seldom falls during that month, and further, that in eleven of the twenty-two years under consideration no rain fell in October. When rain does not fall until the middle or last of November, great anxiety and distress are caused by the delay. In four of the twenty-two years there was a slight fall of rain in September, but rain during this month is to be considered as exceptional.

THE EARLY, MIDDLE, AND LATTER RAINS.

Every one is familiar with the terms the "early" and the "latter" rains, which refer to parts of the rainy season. The rainy season, however, is really divided into three parts, and it is during the middle one of these periods that the most rain falls. It is very seldom that many days of rainy weather occur in succession, but whether the rainy periods are of one or several days' duration, they are sure to be followed by one or many days of fine weather, and these fine days of the Winter and early Spring months are some of the most enjoyable that the climate of Palestine affords. The "early" rains are depended upon to moisten the earth and fit it for the reception of seed, and hence it is a general signal for the commencement of plowing. The middle, or heavy Winter rains, furnish the real water supply of the year. The earth is then saturated, springs are replenished, and cisterns are filled with water. The "latter" rains, which fall in gentle showers, are indispensable to the perfection of the grain. However copious may have been the Winter rains, unless the "latter" rain falls, the harvest is wholly or in part a failure; hence this is looked for by the farmers especially, and by all the people of the land as well, with peculiar anxiety.

CONNECTION OF WIND WITH RAIN.

Most of the rain storms come from a westerly direction. Of those noted during a period of twenty-two years, forty-nine were from the northwest, one hundred and six from the west, and two hundred and thirty-eight were from the southwest. On one hundred and forty-nine occasions, however, an easterly wind immediately preceded the change which ushered in the storm. Not infrequently the direction of the wind changes during the storm; if it passes to the north the rain ceases. A change from any quarter to the southwest usually indicates a continuance of the rain.

SNOW, EARTHQUAKES, TEMPERATURE.

On three hundred and sixty-nine occasions the temperature of the air became lower as the rain fell; on ninety occasions it rose slightly, and on forty-seven occasions it remained stationary, or nearly so, until the rain ceased. During twenty-two years eight seasons have passed without snow, against fourteen seasons when snow has fallen. In general snow falls in small quantities and soon melts, but occasionally there is a heavy fall, that for instance for the twenty-eighth and twenty-ninth of December, 1879, which was extremely heavy, measuring seventeen inches on a level.

Most of the earthquakes that have been noted occurred during the rainy season; eighty occurred during actual storms, and four of these during snow storms. It is interesting to observe further, that in nearly every instance they have been preceded or followed by an easterly wind.

JERUSALEM'S ELEVATION.

The elevation of the city of Jerusalem is 2,600 feet above the level of the Mediterranean Sea, and the mean height of the barometer during twenty-one years was 27.40 inches. The highest reading during this period was 27.82 inches; the lowest reading 26.97 inches, the range being .63.

COLD, HEAT, AND FROST.

The coldest month in Jerusalem is February, the mean temperature during the last twenty-two years being 47.9°. It rises month by month until August, when the mean temperature has been 76°, and then sinks again month by month until the following February; the mean annual temperature during this period being 62.8°. The hottest days in the year do not occur in August, but usually in May, June, and September. The lowest temperature in twenty-two years was twenty-five, on January 20, 1864. A minimum of thirty and thirty-two has been reached in February and October, and once in April. These cases are, however, notable exceptions.

In Jerusalem frost occurs five or six nights in the course of a Winter, but ice is rarely ever formed.

STORMY WINDS.

A peculiar feature of the climate of Palestine is its stormy winds. Its physical conformation has doubtless something to do with this. It is a ridge of rugged mountains, running north and south, which drops to a broad maritime plain on the west, and on the east to a deep chasm sunk into the earth to a depth of thirteen hundred feet below the level of the Mediterranean Sea, beyond which chasm (which is the Jordan River Valley) another ridge of mountains rises abruptly to a height equal to or greater than that of the western ridge, and beyond this, in turn, a vast table land stretches eastward to the Euphrates, and southward into Arabia. Both the inhabitants of the country and its crops are largely affected by the prevailing winds. The north wind is cold, the south warm, the east dry, the west is moist. North and northwesterly winds prevail most in the Summer months, when they are cool and refreshing, moderately dry, and accompanied by a few or no clouds. The north winds of Winter are cold and sharp. Their coolness and sharpness, even in Summer, are apt to produce sore throat, fever, and dysentery.

Without the strong westerly winds of Summer, the climate of Jerusalem would be unbearable. Occasionally these winds do not blow for several days in succession, and at such times the heat becomes very great. As a rule this strong westerly breeze comes up every afternoon. It is felt at Jaffa, and at other places on the coast, as early as nine or ten o'clock in the morning, but it does not reach Jerusalem until from two to four o'clock in the afternoon. Generally it subsides about sunset, but rises again during the evening, and sometimes continues through a greater part of the night. Consequently, however hot the day may have been in Jerusalem, the nights during the Summer season are almost always cool. Thus, this wind, although often strong, disagreeable, and filling the air with clouds of dust, is a great blessing to the inhabitants; but at the same time it makes it very necessary for them to take precautions to protect themselves from its influence at night. Easterly winds are rare in Summer, while they are common in each of the other seasons. The average for sixteen years has been three days of easterly winds for each month from June to September, and eleven days for each month from October to May, inclusive.

EAST WIND AND SIROCCO.

The east wind in Winter is usually accompanied by a clear blue sky, and is dry, stimulating, and if not too strong, very agreeable. In the warmer months it is unpleasant and depressing from its great heat and dryness, and the occasional haze and dust which occasionally accompanies it. The southeast winds are those which are popularly termed "Siroccos," and which are most disagreeable. "The worst kind of sirocco," says Dr. Chaplin, "dries the mucous membrane of the air passages, producing a kind of inflammation resulting in catarrh and sore throat; it induces great lassitude, incapacitating for mental as well as bodily exertion, in those who walk or work in it; headache, with the sense of constriction, as if a cord were tied around the temples, oppression of the chest, burning of the palms of the hands and soles of the feet, accelerated pulse, thirst, and sometimes fever. It dries and cracks furniture, loosening the joints of tables and chairs, curls the covers of books, and pictures hung in frames, parches vegetation, and sometimes withers whole fields of young grain. Its force is not usually great, but sometimes severe storms of wind and fine dust are experienced, the hot air burning like a blast from an oven, and the sand cutting the face of the traveler who has the misfortune to encounter it. This kind of air has a peculiar smell, not unlike that of the neighborhood of a burning brick kiln. Sometimes the most remarkable whirlwinds are produced, especially in the western plain near the hills, by the meeting of a strong east or southeast wind with a wind from the west or north. Clouds of sand fly about in all directions, now taking the traveler in front, now behind, and now on the side, and the gusts of wind are so violent as to blow weak persons from their horses, and overturn baggage animals. The cold siroccos of Winter often blow with much force, and when it comes from a few degrees north of east, it is so cold and piercing as sometimes to kill those who are exposed to it without sufficient clothing, instances of which occurred in 1867."

It is an old and popular saying with the people of the country that a sirocco always lasts three days, but they have been known to last for twenty and even thirty days.

SCARCITY OF CLOUDS.

A noticeable feature of Palestine is its cloudless skies. There has been an average of one hundred and forty days in each year for sixteen years which were cloudless at 9 A. M. Still during a large part of the year, clouds are present, and they affect the climate in various ways, but chiefly by moistening the atmosphere and by producing a shade which moderates the otherwise intense heat of the season. The smallest amount of clouds during the year is in the months of July and August.

DEPOSITION OF DEW.

During the Winter months dew falls in the ordinary way, and hence needs no special notice; but in the Summer months, when the whole country is arid, and there is no water to evaporate, the copious dews are brought entirely by the westerly winds from the sea. If no westerly breeze, or but a slight one springs up towards evening, there is no dew. The heavy dews of Summer, which modify the climate so remarkably, differ from ordinary dew in the manner of their deposition, being in great part precipitated in the air in the form of mist before being deposited on the earth. On Summer evenings a few clouds are commonly to be seen in the western horizon soon after sunset. Later in the evening they increase in number, become lower and looser, and sweep past at no great elevation, and often with considerable velocity. Towards midnight, or later, they become still more abundant and still lower, brushing the tops of the hills as they pass, and depositing much of their moisture upon them, although dew may fall, even in Summer, in the usual way on clear nights; the surest sign of a copious deposition is the appearance of clouds with a westerly wind after sunset. Dew is most copious in the Spring, and in September and October, except during sirocco weather, when there is none.

Clouds and a westerly wind at sunset and afterwards, are not, however, always indications of a very damp night. It is the continuation of the westerly wind during the night that brings abundance of dew. Often at daybreak the sky is obscured by a heavy mist, and the ground is wet as if rain had fallen. When the sun begins to act upon this mist, large masses of white clouds are formed, which, however, soon disappear before the great heat, leaving overhead only the usual blue sky of Summer.

UNHEALTHY PERIOD OF THE YEAR.

The unhealthy period of the year, the period in which the climatic diseases of the country, such as ophthalmia, fevers, and dysentery, are most prevalent, extends from May to October, inclusive. Six things strongly characterize this period: (1) almost entire absence of rain; (2) low atmospheric pressure, with small range; (3) high temperature, with great daily range; (4) great dryness of the atmosphere; (5) a very small amount of cloud; and (6) except at the beginning and end of this period, a minimum of easterly winds. While I have been completing this report the cholera has broken out in Egypt, and a strict quarantine has been established in all the Syrian ports. In this connection, therefore, it may be of interest to add a note respecting the climate of Jerusalem in October, 1865, when the cholera raged in this city with considerable violence. The period from the seventh to the twenty-fourth of that month was one of great and oppressive heat. During the whole eighteen days the sky was cloudless, but overspread with a thin haze. The wind was from northwest, north, and east, but so light as to be considered a calm, except on the fifteenth and sixteenth, when there was a light breeze from the east. The highest temperature was ninety-four degrees, and on eleven days it rose to at least ninety degrees. During the period the mean of the maximum temperature was 89.1, and of the minimum temperature 65.8; the mean temperature for the period being 77.4 degrees. This high temperature, together with the calm, close, hazy atmosphere, was supposed to have some influence in spreading the cholera.

www.ingramcontent.com/pod-product-compliance
Lightning Source LLC
Chambersburg PA
CBHW021952190326
41519CB00009B/1231